T0331671

Math
Makes
Sense!

A Constructivist Approach to the
Teaching and Learning of Mathematics

Math
Makes
Sense!

A Constructivist Approach to the Teaching and Learning of Mathematics

Ana Helvia Quintero
University of Puerto Rico, Puerto Rico

Héctor Rosario
North Carolina School of Science and Mathematics, USA

Imperial College Press

Published by

Imperial College Press
57 Shelton Street
Covent Garden
London WC2H 9HE

Distributed by

World Scientific Publishing Co. Pte. Ltd.
5 Toh Tuck Link, Singapore 596224
USA office: 27 Warren Street, Suite 401-402, Hackensack, NJ 07601
UK office: 57 Shelton Street, Covent Garden, London WC2H 9HE

Library of Congress Cataloging-in-Publication Data
Names: Quintero, Ana Helvia. | Rosario, Héctor.
Title: Math makes sense! : a constructivist approach to the teaching and learning of mathematics /
    Ana Helvia Quintero (University of Puerto Rico, Puerto Rico) &
    Héctor Rosario (North Carolina School of Science and Mathematics, USA).
Description: New Jersey : Imperial College Press, 2016.
Identifiers: LCCN 2015042688| ISBN 9781783268634 (hc : alk. paper) |
    ISBN 9781783268641 (sc : alk. paper)
Subjects: LCSH: Mathematics--Study and teaching.
Classification: LCC QA135.5 .Q55 2016 | DDC 510.71--dc23
LC record available at http://lccn.loc.gov/2015042688

British Library Cataloguing-in-Publication Data
A catalogue record for this book is available from the British Library.

In-house Editors: Chandrima Maitra/Catharina Weijman

Typeset by Stallion Press
Email: enquiries@stallionpress.com

Printed in Singapore

PREFACE[1]

Experience has taught us that a great number of students do not learn the concepts of mathematics that they are taught in elementary school. Nevertheless, these same children use mathematics in their everyday environment. For example, without knowing anything about the formal theory of fractions or rational numbers, children know how to "fairly" divide a pizza or a chocolate bar. This book shows us how to capitalize on this informal knowledge of mathematics that children bring with them, and build upon it the corresponding mathematical concepts in a formal way.

Through examples and activities, the book shows paths for learning with understanding the mathematical concepts taught in elementary school. While discussing these paths, we emphasize that many of the concepts that are taught are much more difficult than some educators might think. Hence, we focus on less content and more depth, based on our teaching philosophy that children should be taught basic concepts with a thorough understanding of what they mean, instead of piling up material with no comprehension.

This book also presents activities that foster the learning of mathematical concepts in an environment of exchange that promotes an interest in solving problems and making conjectures, as well as in sharing, discussing, and

---

[1] This is a revised and expanded translation of the book *Matemáticas con sentido: aprendizaje y enseñanza* by Ana H. Quintero, published by La Editorial Universidad de Puerto Rico. It includes many new illustrations and an additional chapter on mathematical puzzles.

defending ideas with classmates. An environment that encourages student participation — in which students develop basic skills and learn to understand and appreciate the language of mathematics — has as its final goal the promise of enabling students to enter the wonderful world of mathematical creation and imagination.

Yet, regardless of how perfect our presentation of the devolvement of these concepts may be, we will never capture the richness and diversity that are found in the classroom. Therefore, along with the paths we offer, it is necessary to create a school (or home) environment that promotes constant reflection in students and teachers. In this connection, we present ideas in the form of research questions that should prove stimulating for teachers, who are responsible for kindling students' interest.

ACKNOWLEDGMENTS

This book is the product of years of research. Many people have contributed throughout this process and I am grateful to all of them. I especially appreciate the assistance of Dr. Jorge M. López, who introduced me to the field of realistic mathematics education (RME). I have shared with him throughout the years my research as well as my inner thoughts on how mathematics is learned. The CRAIM team (Centros Regionales de Adiestramiento en Instrucción Matemática), led by Jorge, has also been actively involved in this process.

Many teachers from numerous schools have shared their experiences in the classroom with me through active collaboration. Their work has greatly enhanced the experiences presented in this book. In particular, I deeply appreciate the support provided by the teachers and principals at the following elementary schools: Antonio S. Pedreira in Puerto Nuevo, Abraham Lincoln in Old San Juan, Sofía Rexach in Cantera, and Rafael Cordero in Cataño.

I have shared different versions of the various chapters included in this book with Professor Luis López, from the Faculty of Education at the University of Puerto Rico, who used them for his "Methodology of Teaching Mathematics" course. I thank him and his students for their support in the process of clarifying the arguments and explanations presented in those chapters.

The Mathematics Department of the Faculty of Natural Sciences at the Río Piedras Campus of the University of Puerto Rico was always willing to give me the support and time to develop this project. I also thank La Editorial Universidad de Puerto Rico for granting permission to use the illustrations from the Spanish edition. Finally, I appreciate the support for my endeavors given by my family. They shared with me not only the teachings and reflections of this project, but also sustained me emotionally.

AHQ

I would like to express my appreciation for the translating assistance provided by my friend Pedro Zayas (Chapters 2–7). My wife Verónica del Mar and oldest daughter Ishtar also helped me with translating work (Acknowledgements, Preface, and Chapter 8). My family has borne the burden of this rapid translation, revision, and expansion of a book highly regarded by elementary mathematics teachers and teacher trainers in Puerto Rico. I also thank my homeschooled children — Arjuna, Rohini, Vrinda, Haridas, and Krishna — for being my greatest teachers while playing the role of students during my experiments with the teaching of mathematics. My approach with them is sprinkled throughout the book. I am deeply thankful for my family's love and support.

HR

ABOUT THE AUTHORS

Ana Helvia Quintero is Professor of Mathematics at the University of Puerto Rico in Río Piedras and holds a Ph.D. in the learning of mathematics from MIT. Her main research interest is mathematics education and has been involved in several projects aimed at improving the school system in Puerto Rico. From January 2001 to June 2002, she was Under Secretary of Education of the Commonwealth of Puerto Rico. Based on that experience, she published a book in 2006, *Muchas reformas, pocos cambios* (Many Reforms, Hardly Any Changes).

Héctor Rosario holds a Ph.D. in Mathematics education from Columbia University. During the production of this book, he was Professor of Mathematics at the University of Puerto Rico at Mayagüez; he is now a faculty member at the North Carolina School of Science and Mathematics. Dr. Rosario is the principal editor of the 17-nation anthology *Mathematics and Its Teaching in the Southern Americas*, published by World Scientific Publishing. He is also a member of the advisory board of the National Association of Math Circles.

CONTENTS

CHAPTER 1

FOSTERING THE LEARNING OF MATHEMATICS

1.1 Introduction

Mathematics is one of the most difficult subjects for many students. A great deal of this difficulty is a result of how it is taught. The difficulties of learning mathematics can be grouped into two major categories: issues that are similar to those of teaching other subjects — it is taught without understanding and without taking into consideration how it is learned (Bruner, 1990; Cohen *et al.*, 1993; Quintero *et al.*, 2006; Wiske, 1998) — and those that spring from the way in which the teaching of the mathematical concepts is organized. In this book we will discuss both types of issues. However, since the literature is vast concerning the difficulties shared with other subjects, we will only summarize them and recommend some further reading, while emphasizing the problems that spring from how mathematical concepts are organized for teaching. We understand that there is much research to be done on this aspect. In this regard, this book presents the status of current research while intending to promote future studies.

1.2 Problems in the Teaching of Mathematics Shared with Other Disciplines

1.2.1 *Teaching without understanding*

We all remember from our school days some instances when we were taught without understanding. We recall the act of having learned, but not what was taught. Whatever is learned in this way will be forgotten easily and will not contribute to the understanding of our world, either natural or cultural. Every day, research on the teaching for understanding increases (see, for instance, Bruner, 1990; Cohen *et al.*, 1993; Quintero *et al.*, 2006; Wiske, 1998). When learning is relevant, we can apply our knowledge to new situations and make connections among diverse fields of knowledge; we strengthen our competence to use our knowledge (Bransford *et al.*, 2000) and we are able to connect our knowledge to our day-to-day experiences.

Children show interest in mathematical ideas form an early age (McCrink and Wynn, 2004; Whalen *et al.*, 1999). Through their day-to-day experience, children develop informal ideas about numbers, quantities, patterns, shapes, and size, among other concepts. The learning of mathematical ideas begins much earlier than children's formal school experience (Gelman and Gallistel, 1978; Resnick, 1987). Even after entering school, students of all ages develop mathematical ideas on a daily basis (Bransford *et al.*, 2000). A problem with contemporary teaching is that it does not integrate these experiences into the formal classroom learning environment. When one of the authors was working on verbal problems with fifth graders (10–11 years old), she had the following exchange with them after posing this problem:

> *A man bought 20 oranges. If the oranges are 5 for a dollar, how much did he spend?*
>
> *Student: I don't know how to do the problem. I know the answer is $4.00, but I don't know how to do it.*
>
> *Researcher: But, how do you know that the answer is $4.00?*
>
> *Student: Well, if they give 5 oranges for a dollar, for $2.00 they'll give 10, so that 20 oranges will be $4.00.*

We see that students separate two types of knowledge they possess to solve the problem. On the one hand, they have the informal learning of day-to-day life in and out of school. On the other, they carry out the arithmetic operations, which they do not relate to their informal knowledge. They think that solving a problem equals translating it into a single arithmetic operation (addition, subtraction, multiplication, or division), as is demanded from them at school. Being unable to do this translation, they feel that they cannot solve the problem, though they have done so with their informal knowledge.

It is imperative to bridge the informal math that students do daily with the formal math taught in schools. In fact, the math that students learn through their daily interactions is learned with meaning and can be used in diverse contexts, as shown in the example above. The math they learn in school is for the most part a set of meaningless rules that they barely use outside of school. An area of research that college professors and school teachers must share is that of identifying activities that, springing from the students' experiences, support the construction of mathematical concepts.

Mathematics is one of the most misunderstood disciplines. The understanding among a great portion of the population is that mathematics is a series of rules used for numerical calculations. Responding to this interpretation, mathematics is taught as a set of formulas to do various calculations. Mathematics, however, as most fields of knowledge, developed as a result of the human need to understand and interpret the world (Davis and Hersh, 1972; Kline, 1973). The teaching of mathematics must therefore spring from contexts that are meaningful for the students and build on their informal knowledge. For example, if we want to introduce the concept of measurement in second grade (ages seven to eight years), we can start with the question, "How much do we grow in a year?" We start by discussing with students their notions of measurement, and from these build up to the more sophisticated notions that mathematics has developed.

Throughout the book, we present meaningful contexts for teaching various mathematical concepts. Teachers, in turn, will investigate and, based on their research, develop a database of examples of contexts that promote the learning of the mathematical concepts taught. In fact, while teaching most curricular subjects, situations arise that lend themselves to mathematical analysis. For example, when studying the human body in kindergarten, we may ask, "How many eyes?", "How many hands?", or "How many fingers?" Ideally, the teaching of mathematics should be integrated into the teaching of other subjects. If during social studies class we study the community, we can use that context — the corner shop or children in different areas of the community — to present counting problems, as well as addition and subtraction problems (see Fundación Quintero Alfaro, 1993, 1994, for examples of schools in Puerto Rico working with curriculum integration).

1.2.2 *Teaching as information transfer*

Teaching based on meaningful contexts allows students to connect what they are learning with their ideas and experience. But it's not enough just to start from meaningful contexts, it is also important to promote active and constructive learning. The most widespread misconception of learning is that children come into this world as a *tabula rasa*, i.e. their minds contain no ideas or thoughts and that they gain knowledge from their experience and what they receive from adults. In the 1950s, Swiss biologist Jean Piaget (1921, 1966, 1976) presented a different interpretation of how humans learn. He stated that human beings, rather than being processors of information, are creators of models, which they use to understand the world and to theorize about it.

Annette Karmiloff-Smith and Bärbel Inhelder (1975) suggest that children develop theories to explain the world from a very early age. Originally these theories, which they call *theories in action*, are composed of some implicit ideas or representation models of a situation. In their interactions with the physical and cultural world, humans develop their theories. Piaget noted that children did not repeat the explanations presented by adults of an event, but rather built their own from the connections made between their experience and their theories. For instance, until about nine years of age, the model that children use to explain the behavior of living beings and inanimate objects is the model of human behavior in society. Thus, a child before this age will explain the motion of the moon from this model and say, for example, that the moon went for a walk or that it was angry and hid, even after hearing the explanation of the motion of the moon based on physical laws. Children do not merely repeat the explanations offered, but construct their interpretations from their own theories.

Piaget's theory about the process of constructing knowledge has been put forward, independently, by other scholars, such as Vygotsky (1978) and Luria (1976). These theorists emphasize the role of culture in the construction made by the child. From these great thinkers, more recent researchers have developed the constructivist learning theory (Bruner, 1996; Lakoff, 1987).

Constructivism suggests that humans are seeking to understand the world around them. In this quest, they are active entities exploring their environment. During their search, they develop schemes of understanding that expand according to their interactions with their physical and cultural environments. Thus, when confronted with a new experience, they can accommodate it in their current scheme, or revise and extend that scheme to integrate the new experience. For example, the original food scheme of a baby is to suck. At the beginning, babies only drink milk. The first time they take juice, they might seem surprised, but easily accommodate the new experience into their sucking scheme. Now when they are given solid food, they try sucking and make a mess. Gradually, they adapt their scheme and extend it to sucking when the food is liquid and swallowing when it is solid. These processes are what Piaget called accommodation and adaptation. When their mental schemes are not ready to integrate a situation, people may misinterpret the situation by integrating it into one of their existing schemes, or may simply overlook the situation. The example of a child's interpretation of the motion of the moon is an example of the first case.

In the teaching of any discipline it is important to provide the opportunity for students to explore, make conjectures, discuss them, and, gradually, actively build their knowledge. We introduce the context, and in it the question or problem that leads to the subject we want to address. Then we allow students to develop their own solutions. Students can work individually or in groups. Once students have worked on solving the problem, both correct and incorrect strategies are discussed. Through questions, we promote students' learning from each other. Allowing students to build their solutions enables them to build their own mathematical knowledge. This classroom dynamic provides an opportunity for the teacher to identify the students' concepts and ideas, and thus support their construction upon their conceptions. For instance, let us assume that we are in the first few weeks of school. In fifth grade (10–11 years), we are interested in developing the division algorithm. In the first few days, we meet with parents and take advantage of this to present the following problem:

> *81 people will be present at the Parent Teacher Association meeting. If we can accommodate 6 people per table, how many tables do we need?*

In a study conducted in a classroom we found the following strategies for solving this problem:

— Add $6 + 6 + 6 + \cdots$
— Count 6, 12, 18, ...
— Start from 81 and subtract 6 many times
— Say the times table $1 \times 6, 2 \times 6, 3 \times 6, \ldots$ until we reach the product closest to 81.

An analysis of these responses shows that the child who adds $6 + 6 + 6$, etc., does not interpret the above problem in a multiplicative structure. Prior to introducing division we have to develop the idea of multiplication. The work of the fourth child, meanwhile, shows that he sees the multiplicative structure of the problem. In this case, we can introduce the concept of division based on his knowledge of multiplication.

We then introduce other problems with the structure of division, for example:

> *At the meeting we want to give parents soda. We decide on 2-L bottles. If each bottle serves 8 people, how many do we need to buy?*

After several problems with the same structure, we invite students to identify the common elements in them. From those methods they develop the algorithm. In Chapter 5, we will discuss the cognitive development of division and how to go about building the algorithm.

The process we have discussed in the example of the division problem must guide us in the conceptualization of mathematical terms. Thus, by introducing a problem or situation, we allow students to create their own strategies. This helps students to build their knowledge from their informal knowledge. Moreover, it allows teachers to create activities based on students' conceptions, which will serve to broaden, deepen, or correct them. Therefore, it is important for teachers to be attentive to the students' output, hear and observe their explanations, and foster reflection on them. In this process, the teacher directs students in analyzing their conceptions (Lampert 1989; Mack, 1990). Teachers should give room for students to discuss among themselves their strategies, thus learning from others as everyone explains their reasoning.

The active and constructive teaching process takes longer than the transfer of knowledge method. It is much easier to explain the division algorithm than it is to assist students in the construction of the elements that comprise it. Ultimately, however, the time saved is not worth it because students do not really learn, and even though they can pass an exam by mechanically applying information they have learned, that information will be easily forgotten because there was no integration into their conceptions and the material must be retaught. That is why we still find college students who do not understand fractions.

1.2.3 *Teaching without reflection*

In mathematics, concepts arise from reflecting upon mathematical activity. When we ponder different situations, we discover similarities and common elements that are abstract paradigms and developed into mental action patterns. Such patterns, similarities, and common elements are organized and systematized in mathematical structures. Once students have considered several situations that integrate a mathematical concept, they can abstract the mathematical structure through reflection. For example, once students have worked several problems with the same structure, these are discussed so as to identify common elements. At this point, symbolism, algorithms, and mathematical structures begin to be developed. For example, after working several problems with the structure of division — as those presented in the previous section — students realize that to find how many groups of the same quantity fit into a number (division model), it is more efficient to multiply than to repeatedly add the number of objects in the group. They can then find the major product of the divisor that fits into the dividend. From this pattern, a scheme emerges that includes the students' strategies, which in turn approaches the division algorithm. We see that the abstraction process should be gradual. We start with a concrete situation, which varies according to the level. For example, for students analyzing division problems for the first time, the algorithm of this operation is abstract. Once the student has experience with division, they can use it in the construction of other operations.

From the analysis of "concrete" situations, students create their own nomenclature to represent them. This is the pre-formal stage in which students introduce symbols, drawings, and diagrams that help them in the qualitative analysis of the situation. From the symbols, algorithms, strategies, and models that students make, we develop the mathematical language and symbols. This process takes longer than if we simply present and explain mathematical symbols. In the long run, however, we recover this time as students actually learn and we don't have to be in endless remedial sessions.

Hence, in each level we follow this pattern:

concrete → pre-formal → formal

For instance, while developing subtraction we have:

Concrete

Mary has 5 lollipops and gives 2 to Lilly. How many is she left with?

Pre-formal

Mary gave away 2 lollipops.

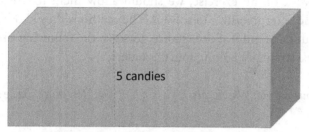

5 candies

How many candies are in the second section?

Formal

$5 - 2 = 3$.

The concrete also varies with the level of student development. When learning addition, the "concrete" is counting. Once the student has learned and automated addition, addition becomes "concrete" to work a multiplication or division problem. In this way, the concrete is not limited to physical materials or real situations. What is concrete is rather a situation or problem that students can interpret with the knowledge they have. In fact, for an advanced student of mathematics, algebra is a "concrete" example of a mathematical structure. Skipping directly to mathematical abstractions encourages students to see mathematics as a process that is foreign to intuition and experience.

Reflection should be encouraged both for individual work and group discussions. In the latter, teacher intervention should drive the analysis and research instead of presenting the answer. Using the question-and-answer method, we guide students toward their awareness of the effectiveness of certain strategies.

1.2.4 *Homogeneous instruction*

For decades, it has been suggested that all students are different and follow different routes in their learning. However, this principle with which everyone agrees is not taken into account in curriculum development and teaching. We must recognize that the development of knowledge does not occur in the same way in all people (Gardner, 1983). For example, when students work on an exercise, we should allow them to use their own method and interpretation to solve it. The methods they choose show us how they understand it. For instance, to solve the following problem we found three strategies among fourth graders.

Rick bought 6 Hot Wheels cars. Each cost him 35¢. How much did he spend?

Strategy 1: Counting

The student draws sticks to represent the pennies that each car cost and then counts them one by one.

||||||||||||||||||||||||||||||||||| |||||||||||||||||||||||||||||||||||||||

||||||||||||||||||||||||||||||||||| |||||||||||||||||||||||||||||||||||||||

||||||||||||||||||||||||||||||||||| |||||||||||||||||||||||||||||||||||||||

Strategy 2: Addition

$$
\begin{array}{r}
35 \\
35 \\
35 \\
35 \\
35 \\
+\ 35 \\
\hline
\end{array}
$$

Strategy 3: Multiplication

$$
\begin{array}{r}
35 \\
\times\ 6 \\
\hline
\end{array}
$$

Every method is correct, but some are more efficient than others. Each strategy allows us to see how the student interprets the problem. The fact that the children who employed the first strategy did not use addition or multiplication to solve this problem says that they do not understand what these operations exemplify. They might be able to solve addition or multiplication exercises but do not understand what those operations mean. In this case, it is necessary for them to construct the meaning of these operations.

Knowing how students interpret and solve different problems gives us an idea of the diversity in levels of analysis that coexist in the classroom; it is a way to assess through discussion. Therefore, it is very important to observe students and try to comprehend how they are understanding the situation. Only then can we help them in the construction of their knowledge.

We currently have students with different learning abilities. For teachers, this presents a challenge because they either bore some or lose others. This area — differentiating instruction — is a relatively new concept and requires further research (Small, 2012). It aims at answering the question of how to create environments that address diversity. For this, it is important to pre-pare activities that lend themselves to different learning styles and rates (Small, 2012). We suggest that once an activity is presented, time be given for group discussion working in small groups sorted by aptitude level.

Another way to address diversity is to use more advanced students as tutors. It has been found that both the mentor and the mentee benefit from this experience (Elbers, 2003). The tutors, by having to explain con-cepts, will deepen their understanding.

1.2.5 *Teaching as an individual process*

In the construction of knowledge, sharing with members of our commu-nity is of great importance. One criticism that has been made against Piaget's theory is that it presents learning primarily as an individual pro-cess. Bruner (1996), for example, has argued that it is not a solitary task, but that it occurs in a society and is directed and encouraged by the socio-cultural context. Therefore, the environment that we develop in the class-room is very important: it must foster an exchange among learners. In this connection, Elbers (2003) presents a lesson in which the class becomes a group that asks and discusses, and students learn from each other. The class begins with the introduction of a problem. Students work the problem individually, but with the expectation that they have to present and justify their solution to the group. As students work individually, the teacher observes students whose solutions show an ingenious method to solve the problem. The teacher asks them to share their answers in front of the class. The discussion of ingenious strategies invites the group to reflect on their strategies and to learn more innovative ones.

We must be aware that it is a mistake to equate education with schooling and teaching because in reality these are only two components, among many others. Education is a process that begins at birth and occurs in multiple scenarios, such as in the family, in the community, and in the media. In school education, we must be aware that students bring many experiences, ideas, and information from other contexts. It is important to build on this knowledge previously acquired in the home and outside of school.

We teach not only by what we say but by what we do. The dynamic that develops in the classroom carries a message of how mathematics is constructed. For instance, an important rule to observe during classroom interactions is to listen to the strategies and justifications offered by each student, and these should be taken seriously when trying to validate a result, either to confirm them or to refute them.

1.2.6 *Mathematics is a cold discipline*

The first author remembers a teacher during the learning of the mathematics research process who wondered how it was possible for students to enjoy coming with her, "if mathematics is so boring." The first author thought that if math was boring to the teacher, it would be very difficult to make it interesting for her students. Although it might seem contradictory or redundant, it is necessary to promote the interest of prospective math teachers in the subject they intend to teach. As we present the activities to develop different mathematical concepts, we will try to show the beauty of and create interest in these concepts. We will also use games as a way to enliven mathematics classes.

1.3 Conclusion

The teaching of mathematics should consider the aforementioned learning principles:

(1) **Teaching for understanding:** In order to achieve understanding, it is imperative to bridge the informal mathematics that students do daily in context with the formal mathematics taught in schools. From Chapter 3 on, we will present situations in contexts that teachers can include in their teaching practice to help students understand the mathematical concepts.

(2) **Teaching as a construction process:** It is not enough to start from meaningful contexts; it is also important to promote active and constructive learning. In Chapter 2 we will discuss how mathematical concepts are constructed in general. From Chapter 3 on, we will discuss students' learning strands of the basic mathematics concepts taught at the elementary level.

(3) **Integrating reflection:** In mathematics, concepts arise from reflecting upon mathematical activity. When we ponder different situations, we discover similarities, common elements, and abstract paradigms into mental action patterns. Hence, reflection is basic for mathematics learning.

(4) **Differentiating instruction:** It is important to prepare activities that lend themselves to different learning styles and rates. We suggest that once an activity is presented, time be given for group discussion working in small groups sorted by aptitude level. Also, students with good verbal skills but weaker mathematical skills can be paired with students who have good mathematical skills but poorer verbal skills. That way students can draw on the strengths of their peers.

(5) **Learning as a social enterprise:** Learning occurs in a society and is directed and encouraged by the socio-cultural context. Therefore, the environment that we develop in the classroom is very important; the environment must foster exchange among learners.

(6) **The joy of learning mathematics:** As we present, through the different chapters, the activities to develop different mathematical concepts, we will show the beauty of and create interest in these concepts. We will also use games as a way to enliven the mathematics classroom.

However, even when following these principles there is no single recipe for implementing them, since the school reality has a multiplicity of variables — including student diversity — that intertwine. These must be addressed in order to develop a curriculum. In this sense, it is essential for teachers to maintain an inquisitive attitude, to always be willing to learn from their practice, and to have the initiative to research and develop alternatives according to their reality.

References

Bransford, J.D., Brown, A.L. and Cocking, R.R. (eds.). 2000. How People Learn. Washington, D.C.: National Academy Press.

Bruner, J. 1990. The Act of Meaning. Cambridge, Mass.: Harvard University Press.

Bruner, J. 1996. The Culture of Education. Cambridge, Mass.: Harvard University Press.

Cohen, D.K., McLaughlin, M.W. and Talbert, J.E. (eds.). 1993. Teaching for Understanding. San Francisco: Jossey-Bass Publishers.

Davis, P.D. and Hersh, R. 1999. The Mathematical Experience. New York: Houghton and Mifflin, Company.

Elbers, E. 2003. "Classroom interaction as reflection: Learning and teaching mathematics in a community of inquiry", Educational Studies in Mathematics, 54, 77–99.

Fundación Ángel G. Quintero Alfaro. 1993. Cuaderno 1: Premio a la Escuela Antonio S. Pedreira. Hato Rey: Comunicadora Nexus.

Fundación Ángel G. Quintero Alfaro. 1994. Cuaderno 2: Que no se quede nadie sin aprender. Premio a la Escuela Juan Ponce de León. Hato Rey: Comunicadora Nexus.

Gardner, H. 1983. Frames of Mind: The Theory of Multiple Intelligence. New York: Basics Books.

Gelman, R. and Gallistel, C.R. 1978. The Child's Understanding of Number. Cambridge, Mass.: Harvard University Press.

Karmiloff-Smith, A. and Inhelder, B. 1975. "If you want to get ahead, get a theory", Cognition, 3, 199–212.

Kline, M. 1972. Mathematical Thought from Ancient to Modern Times. New York: Oxford University Press.

Lakoff, G. 1987. Women, Fire and Dangerous Things: What Categories Reveal About the Mind. Chicago: University of Chicago Press.

Lampert, M. 1989. "Arithmetic as problem solving", Arithmetic Teacher, 36, 34–36.

Luria, A.R. 1976. Cognitive Development: Its Cultural and Social Foundations. Cambridge, Mass.: Harvard University Press.

Mack, N.K. 1990. "Learning fractions with understanding: Building on informal knowledge", Journal for Research in Mathematics Education, 21, 16–32.

McCrink, K. and Wynn, K. 2004. "Large number addition and subtraction by 9-month –old infants", Psychological Science, 15, 776–781.

Piaget, J. 1921. "Essai sur quelques aspects du développement de la notion de partie chez l'enfant", Journal de Psychologie Normale et Pathologique, 18, 449–480.

Piaget, J. 1966. The Origin of Intelligence in Children. New York: International Universities Press.

Piaget, J. 1976. Judgment and Reasoning in the Child. Towata, N.J.: Littlefield, Adams and Co.

Quintero, A.H., Molina, A., Piñero, E., *et al.* 2006. Educación con Sentido: La Educación Ideal y Posible. San Juan: Publicaciones Puertorriqueñas.

Resnick, L.B. 1987. Education and Learning to Think. Washington, D.C.: National Academies Press.

Small, M. 2012. Great Ways to Differentiate Mathematics Instruction. New York: Teachers College Press.

Vygotsky, L. 1978. Mind and Society. Cambridge, Mass.: Harvard University Press.

Whalen, J., Gallistel, C.R. and Gelman, R. 1999. "Nonverbal counting in humans: The psycho-physics of number representation", Psychological Science, 10, 130–137.

Wiske, M.S. 1998. Teaching for Understanding: Linking Research with Practice. San Francisco: Jossey-Bass Publishers.

CHAPTER 2

CONSTRUCTION OF CONCEPTS AND MATHEMATICAL INTERPRETATIONS

2.1 Introduction

In the previous chapter, some essential elements for the construction of mathematical concepts were presented. As Bruner (1996) says, the process of construction of knowledge depends, in part, upon the nature of the task. Therefore, to support the elaboration of mathematical knowledge, along with the principles of learning common to other fields, it is necessary to recognize the particulars of learning mathematics.

Mathematics is characterized by its abstract and formal nature. Studying the history of mathematical abstractions shows us that they arise from human reflection over human actions (Davis and Hersh, 1981; Wilder, 1973). In this process, similarities, common elements, and paradigms are discovered that are abstracted into mental activity patterns. These patterns, similarities, and common elements are then organized and systemized in mathematical structures. In other words, when considering mathematical problems, invariants are identified in different situations from which mathematical structures arise. For example, once you have solved a few problems by counting, then you observe certain patterns that let you simplify counting, giving birth to addition. Step by step, human beings become so familiar with addition that the operation becomes second nature, almost automatic. Addition acts as the basis on which to develop the concept of multiplication. In fact, if we study the origins of the most abstract concepts of mathematics, we see that these originally arise from the analysis of commonplace situations.

The child, through experience, constructs mathematical concepts by reflection over actions. Teaching should encourage this process. To achieve this, the concepts must be elaborated following the logic of learning.

2.2 Development by Following the Logic of Learning

In the process of constructing mathematical concepts, it is important to consider that the logical order in which mathematics presents them is not the natural way of learning them. The mathematics we receive is the product of thousands of years of evolution. The order in which these concepts are presented is not the same order in which they were discovered or created — as, for example, in geometry. Human beings had worked on and studied triangles for centuries before conceptualizing the idea of a point, a line, and a plane. Once the concepts of point, line, and plane were in hand, triangles came to be defined from them. The presentation of the discipline follows the logical order: first we have analytically simpler terms, i.e. terms that serve as basic definitions — in the case of geometry the concepts of point, line, and plane — and then concepts that are defined from them, like triangles, angles, and circles.

However, the logic of learning is different. For students, it is easier to understand the idea of a triangle rather than that of a point, line, or plane, because they have more informal experience with triangles than with points, lines, or planes. These concepts are mathematical constructions; they are more abstract. **Thus, the logic of learning is not the same as the logic of the discipline under study.** In fact, the logic of learning is more related to history than to the development of concepts of the discipline. Studying the evolution of the concepts sheds light on the study of the conceptual development in children (Carey, 1985). One of the objectives of this book is to present what is known as the "logic of learning" about the mathematical concepts taught in elementary school. In other words, what is the natural order of learning them? This knowledge will help support those in charge of building the curriculum to order the presentation of concepts, as well as teachers to help students in the acquisition of their mathematical knowledge. It will also be useful for those interested in research on the cognitive development of mathematical concepts that have not yet been studied. In fact, research topics will be suggested throughout the text.

2.2.1 *The cognitive development of concepts*

At least two elements coincide to determine the logic of learning: the cognitive development of the concepts and their internal organization in mathematical structures. The cognitive development of the concepts refers to the order in which they are formed. Research in children shows that, in many instances, this order is not in parallel with the order in which these concepts are presented in the curriculum (van den Heuvel-Panhuizen, 2003). The latter order follows the logic of the discipline. Here's an example.

Example 1 — adding fractions and equivalent fractions

Adding fractions, when they are heterogeneous, requires expressing them in equivalent fractions. When teaching, the logic of the discipline is the following:

Equivalent fractions → Addition of heterogeneous fractions

But children have a better understanding of the idea of addition than that of equivalent fractions.

Thus the cognitive development of the concepts is the following:

Intuitive addition of heterogeneous fractions → Necessity of equivalent fractions → Equivalent fractions → Addition of heterogeneous fractions

Let us look at a lesson in which we teach the concepts according to the cognitive development of the concepts:

Jorge, Mari, and Laura are camping. Laura loves cooking and she has brought her cookbook. However, she forgot her measuring cups. Mari came up with the idea of using some empty cans by marking different fractions. Here are the cans and the fraction that they want to mark on each one. Help them mark the cans:

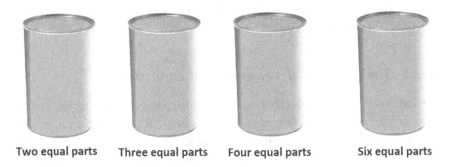

Two equal parts   Three equal parts   Four equal parts   Six equal parts

A recipe asks for 1/2 cup of orange juice and 3/4 cups of water. Jorge says they can pour both liquids in the same can. Mari says it cannot be done because it would spill. Who is correct? Why?

Mari persuades Jorge by drawing the following sketch:

He notices that 1/2 is more than 1/4. So when the can is 3/4 full, there is only 1/4 cup remaining empty, and pouring the 1/2 cup would spill the orange juice.

Laura wants to accommodate the two liquids in two cans, but she wants to fill up one can. How much liquid would be left for the second can?

(There are several ways to solve this problem. In most cases it will be evident that 1/2 = 2/4, leading to the idea of equivalent fractions. However, there are other ways to solve this problem without the need to find equivalent fractions formally.)

While camping they discover some coconut trees. Jorge manages to cut down two coconuts. He extracts their water and pours the water from each coconut into a different can. One of the coconuts fills 1/2 of the can, and the other one fills 1/3 of another can. Is there enough room in one can to pour the water from both coconuts? Why?

How much coconut water would there be in both cans if they were joined?

(The solution to this problem leads to the necessity of equivalent fractions.)

Analyzing these two situations, we observe that to add heterogeneous fractions it is necessary to express them with a common denominator, which, in part, requires equivalent fractions.

This example shows us that keeping in mind the cognitive development of concepts is a must when creating a curriculum. By presenting situations, we promote the construction of those concepts, which, as we saw in this example, do not necessarily follow the logic of the discipline.

One of the basic tasks for research is to discover the cognitive development of the concepts. It is also necessary to research how teaching can be transformed to promote learning following this development.

Besides reconsidering the order of presentation of the concepts, it is necessary to reevaluate the idea that when introducing a concept we must define it in a clear and precise manner. Research shows that development of the concepts occurs at different levels of comprehension. In fact, this process is continual in the student. A concept such as the square shape, for example, is more deeply analyzed as the person develops different levels of knowledge. A child in kindergarten can recognize a square by its shape and distinguish it from a rectangle, parallelogram, or rhombus. At this stage, these concepts are different for the child and none of them includes the others, since their pictorial representations are different. Then, the child starts to study the properties of these figures and starts seeing some likeness among them. For example:

- They all have four sides.
- They all have parallel opposite sides.

Yet, each has its own characteristic traits.

When students start analyzing the relationships between properties, they see, for example, that the parallelogram (a quadrilateral with opposite sides parallel and equal) is a wider concept that includes rectangles as well as squares. In this manner, they can begin to construct conceptual maps in which the relations between these figures are noticed:

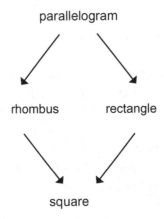

It is therefore important to allow students to explore and analyze the same ideas at different developmental stages, so that the ideas ripen and the students' understanding deepens. Teaching should present a concept at different levels: first in intuitive form, then gradually analyzing its properties until a definition is constructed. In this process it is necessary to keep in mind the research from van Hiele (1986), who argues that learning moves at different levels in the diverse areas of knowledge. It is important to note that these levels are not homogeneous for everyone, but that one person could be at different levels in the various areas of knowledge. We continue with a presentation of these levels using geometry as a context.

Van Hiele Levels

Level 0   **Visualization:** Objects are the basis of knowledge. In the case of geometry, identifying geometric figures by their shape.

Level 1   **Analysis:** The focus of knowledge is the properties of the basic objects. For instance, observing that all sides of a square are equal.

Level 2   **Abstraction:** The focus of knowledge is the relationship among different properties. For example, observing that all squares are also rectangles.

Level 3   **Deduction:** The focus of knowledge is the sequence of sentences, like proofs, where implications and other relationships among propositions are studied.

Level 4   **Rigor:** The focus of knowledge is deductive systems. For this, different axiomatic systems are compared and analyzed, like Euclidean and non-Euclidean geometries.

The van Hiele levels explain some of the communication problems that arise in teaching mathematics. For example, when speaking of the same term, teacher and student could be referring to the same concept, but at different levels. Thus, the phrase "this is a rhombus" may have a different meaning for the student at a certain level from that understood by the teacher. The student recognizes the form and associates it with the word "rhombus." For the student, it is not evident that a square is a rhombus, because their shapes are different:

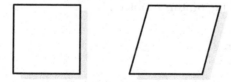

For mathematicians, and for teachers, the concept of rhombus has a different meaning. For them the word "rhombus" means a set of properties and relations: a parallelogram with all sides congruent.

The teacher calls a square a rhombus owing to its properties. The student who is focused on shapes does not associate the square with the rhombus. Van Hiele argues that both have a concept of rhombus, but at different levels.

2.2.2 *The internal organization of concepts in mathematical structures*

The second element that must be analyzed to develop teaching that follows the logic of learning is the internal organization of the concepts in mathematical structures. When considering problems in mathematics, invariants are identified in different situations, from which mathematical structures arise. As with concepts, it is important to support the constructive process of the development of structures. Freudenthal (1991) identifies two processes rolled into one: horizontal mathematics and vertical mathematics. In horizontal mathematics, the student analyzes richly detailed situations, which must be organized and analyzed using mathematical ideas. When working with situations, representations should help visualize concepts and discover patterns and relations; to model the situation with readily known structures, to experiment, classify, and demonstrate.

For example, we want to develop the concept of division starting with the following problem:

> *The PTA meeting expects to sit 81 people. If we can accommodate 6 people per table, how many tables do we need?*

We allow students to do their own analysis and use the strategy they deem most appropriate. As noted in Chapter 1, we found the following strategies for working on this problem:

Add $6 + 6 + 6 + \cdots$
Count 6, 12, 18, ...
Start from 81 and subtract 6 repeatedly
Recite the multiplication table $1 \times 6, 2 \times 6, 3 \times 6, \ldots$ up to the multiplication nearest to 81.

For the last strategy, say $6 \times 10 = 60$, thus if I have 10 tables, then 21 people will be left without a seat; $6 \times 3 = 18$, so that is 3 more tables and 3 people left to sit. So, it is 13 tables plus 1 more table with 3 people.

The various strategies are discussed so that students become familiar with the most efficient ones. In fact, the one used by the last girl is the basis of the division algorithm.

At this stage, we are in **horizontal mathematics**. Students have been exploring and analyzing one situation.

Other problems with divisional structures will be presented, for example, "You want to have some beverage available for the parents' meeting. You decide on 2L bottles of soda. If each bottle can serve 8 people, how many bottles do you need?" Students are allowed to explore again. This stage is also part of horizontal mathematics.

After several problems with the same structure, students are invited to reflect on identifying the common elements found in division problems. From the methods that students develop, algorithms are also developed. In Chapter 5, in which we will discuss the cognitive development of division, we analyze how to go about building the algorithm. At this stage we use vertical mathematics. We are identifying common elements in various situations that lead us to construct mathematical concepts and processes.

In both horizontal and vertical mathematics we should be aware that the focus of humans when analyzing reality is not the pursuit of information, but the need to interpret and understand it. In this process, human beings are not limited to analyzing experiences through simpler concepts, but also look for similarities with previous experiences. Metaphor and metonymy will be important tools to interpret a situation (Lakoff, 1987). This had already been recognized by Thomas Kuhn in *The Essential Tension* (1977). Kuhn states:

> Lacking time to multiply examples, I suggest that an acquired ability to see resemblances between apparently disparate problems plays in the sciences a significant part of the role usually attributed to correspondence rules. Once a new problem is seen to be analogous to a problem previously solved, both an appropriate formalism an a new way of attaching its symbolic consequences to nature follows. [...] Acquiring an

arsenal of exemplars, just as much as learning symbolic generalizations, is integral to the process by which a student gains access to the cognitive achievements of his disciplinary group.

A task of teaching is to familiarize students with the types of situations that can be rendered as metaphors that allow students to see the relationships between them, and from which mathematical structures arise. Thus, we can see the importance of analyzing real situations as a starting point to develop meaningful mathematical structures. However, we must clarify that real events are not necessarily limited to daily life. They may happen within mathematics itself. The key is that the relationship be meaningful to the student. In fact, as students become familiar with numbers, these become contexts for studying relationships and patterns.

Once we introduce a meaningful situation, we let students analyze, draw diagrams, or offer their explanations. Mathematical methods and symbols are constructed from what the students produce. **Vertical mathematics** is the process of reorganizing these mathematical ideas (Freudenthal, 1991; Gravemeijer, 1994; Streefland, 1991; Treffers, 1987). We look for ways to shorten the procedures, connections between concepts, and ways of organizing them.

Supporting the processes of horizontal mathematics and vertical mathematics requires developing sequences of activities that allow students to explore and, through reflection on the regularities, abstract mathematical structures. In this sense, solving activities with isolated problems does not help to develop mathematical structures. It is necessary to develop an articulated sequence and follow the "logic of learning." Let's see an example.

Example 2 — development of the structure of our numerical system

Our numerical system is one of the most important achievements of humanity. It presents an impressive organization and structure. It is not found in the initial forms that humans used to express numbers, but rather displays depth in mathematical development. Understanding this system is not easy, and it also requires a mature understanding of number as a concept. Thus, it is necessary to be aware of some of the properties of numbers, and the relationships between these properties. In this sense, we should not rush to introduce the notion of tens, hundreds, etc., until the student begins to understand the concepts on which these notions are mounted: the idea of part–whole, the idea of each place representing a different type of object — that is, groups of powers of ten. Let's see a curricular sequence that facilitates structuring our numerical system.

The part–whole idea

Our system is based on being able to express a given quantity as the sum of other quantities. Thus, 245 = 200 + 40 + 5. From first grade (six to seven years old), we have to begin forming the idea that a number can be broken down into other numbers. At the end of first grade, we start introducing exercises in which this breakdown uses the number ten as one of those parts.

Horizontal mathematics

The following problem is presented for group analysis:

> *85 students go on a school field trip. We will rent some buses for student transportation. Each bus can accommodate 10 people and all buses must be filled. If a bus is not full, a car will be rented for the remaining students. How many buses must we rent? Do we have to rent a car? How many children will go in the car?*

We must allow students to solve the problem with the method best understood by them. Once they solve it, we discuss the different strategies. For example, when working this problem with fifth graders (10–11 years), I found that almost none of them directly associated how the number is expressed — 8 tens and 5 ones — with the solution of the problem. But these same students could correctly identify the tens and units in a given number.

Most students used toothpicks to form groups of 10 until they reached 80. Others began with 85 sticks and then formed groups of 10.

To wrap up the discussion, let us note the similarity between the result and the way the number is written.

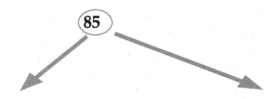

8 buses (with 10 students each) and 5 students in a car

Vertical mathematics

After several problems, we ask students: "Is the resemblance between the result and the way the number is written pure chance?" This is the beginning of analyzing the relationships between the concepts that support the development of the structure of our number system.

In fact, children do not naturally organize numbers in groups of ten. This became clear in an activity by Mrs. Gerena, from Antonio S. Pedreira elementary school, in her second-grade classroom (students of seven to eight years old). She asked the children to divide 17 into different groups.

Different children gave different answers:

$$17 = 16 + 1,$$
$$17 = 8 + 9,$$
$$17 = 12 + 5.$$

But no child in the room said 10 + 7. It is important to introduce activities that require division into groups of ten and then reflect on the groups that arise and on notation for that number, as stated above.

Counting by tens

At the same time that we visualize the idea of dividing a number into groups of tens and ones, we start creating the sequence of counting by tens. Beginning in second grade (seven to eight years), we introduce activities that develop counting by tens on the number line.

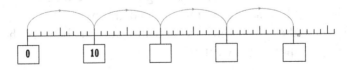

Note that counting by tens is promoted not just using tens as starting points, but from any number because it is important for students to discover that 24 + 10 = 34, 34 + 10 = 44, and so on.

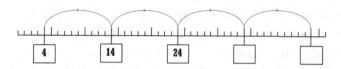

This property that we see as obvious is not so for students. It is necessary to offer them an opportunity to discover it, not just simply hand it to them.

In third grade (eight to nine years), while other decimal places are analyzed, problems that develop counting by hundreds are introduced.

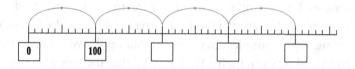

Each place represents a different kind of object

Another important feature of our system is that it is positional: each place represents a different unit, which in our case is base ten. The understanding of this idea is not grasped until the end of elementary school. However, we can start introducing activities that promote the development of these ideas in primary grades.

Integrating metaphor: Coins

Our monetary system also uses a decimal system: ten cents make a dime, ten dimes make a dollar. From the second grade (seven to eight years) on, we introduce activities analogous to the monetary system that support comprehension of the number system.

Once the student has experience with these representations we can introduce the Cuisenaire rods, which already include the model of our system, originally associated with coins.

Concrete materials can help students grasp the structure of our number system, but the teacher must make sure that students do not get used to working with them mechanically, without reaching for an understanding of the structure. This is why teachers' questions leading to reflection are so important.

Finally, around third grade (eight to nine years), we begin teaching the notation of ones, tens, and hundreds.

We propose a different curriculum order from the present one. We start by exploring numbers, observing their organization in groups of ten. It is not until the student has this experience and an understanding of the part–whole property that we introduce the properties of our system. Initially we use metaphors (pre-formal) to analyze these properties: groups of straws, the monetary system, the abacus. We suggest not formalizing the idea of ones–tens–hundreds until third grade.

In the process of supporting students to build mathematical structures, we must evaluate some materials that are presented as constructivist. Sometimes, we find examples of mathematical materials and activities that supposedly follow the constructivist approach, yet they do not promote real learning. For example, they present manipulatives for students to build abstract ideas from them. They assume that the task of constructing means going from the concrete — understanding the concrete object as either a real situation or the manipulatives themselves — to the abstract. They are not aware that this process takes different routes according to the subject under study. In the case of mathematics, many of the manipulatives offer an already-made mathematical structure, although in concrete form. Thus, the opportunity for students to build that structure is missing.

For example, when teaching our number system the starting point is the Cuisenaire rods:

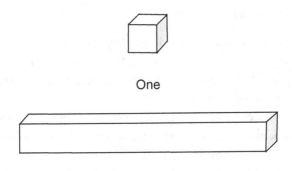

One

Ten

When children work with Cuisenaire rods before understanding our system, they do not see the structure of our system in the rods, but rather see a series of blocks without making the connection with numbers. One of the authors once used Cuisenaire rods with a group of first graders (six to seven years). The task consisted in representing numbers with blocks. She asked them to represent 5, 12, 16, 18, 24, and 35. Before requesting a representation, she showed them how ten unit blocks correspond to a block of ten. Every child began to represent numbers with units. When they were on 16 she said, "But notice that you can simplify the task if ten is represented by this block [the ten]." Some children did not understand her and carried on with units. Those who understood the suggestion used the ten block when they reached 18. However, they represented number 24 with a ten block and 14 units. They had seen 16 written as 10 plus 6 and they carried on likewise with 18, apparently understanding the system. When they represented 24 as 10 plus 14, she realized that the children had followed my suggestion to change ten units for a ten block, but that they did not understand the system. The blocks were not helping much in this task. Indeed, during that research we realized that learning our number system is much more complex than we think.

Cuisenaire rods embody the structure of our system: ones, tens, hundreds, and thousands. The student is expected, by abstraction, to develop the number system with the concrete model as a starting point. This approach does not take into account what we explained about the development of mathematics. In these manipulatives, the concepts and structures are present in the concrete materials. However, if students are not involved in the process of building these structures, then they will not recognize the structure in the materials.

In many cases the structures that the manipulatives represent are clear for the teacher, but not for the student. Because students do not understand them, all they see in the materials are just materials. They are not offered an opportunity for the required understanding process. We should clarify that this does not mean that we should do away completely with manipulatives. What it does imply is that manipulatives, without the process of construction and reflection, are not very helpful. Yet, if we use the materials as building tools, they can be valuable. For example, once the student has experienced a curricular sequence as previously described and developed the idea of ones, tens, and hundreds, we can use blocks to explain rules of operations.

2.3 Slow and Steady Wins the Race

This popular saying also applies to the teaching of mathematics. Students must construct their own knowledge. We start with a concrete situation, which varies according to the expected level of abstraction of students. For example, numbers are abstract for children in the early grades. Once students have substantial experience with numbers, these become concrete examples of algebraic expressions. However, we should note that the path of the teaching–learning process may vary from one concept to another. For example, a student may be at level 2 in the development of numbers and at level 3 in the development of geometric figures.

Starting from the analysis of "concrete" situations, students develop their own nomenclature to represent the situation. This is the pre-formal stage in which the students begin to use symbols, drawings, patterns, and strategies to help them analyze the situation. Based on the symbols, algorithms, strategies, and models that the students create, we build mathematical language and symbols. We should be aware that concept development transits at various levels of abstraction, as we saw in the discussion of van Hiele's levels. At each level we must allow students to construct their own knowledge. This process takes more time than if we simply present and explain mathematical symbols. However, this time will be recovered as students will have learned with understanding, thus avoiding endless remedial sessions.

2.4 Final Reflections

The reflection about learning that we have presented in this first part of the book identifies a set of elements that are very important and must be considered in the teaching of mathematics for understanding:

- The teaching of concepts with meaning.
- The development of higher-order thinking skills, such as reflection.
- The development of mathematical enjoyment.
- The importance of transferring knowledge among different contexts.

Considering the cognitive developmental theories related to these elements, the following teaching strategy arises as will be presented below, but first keep in mind that we only present an outline and that every teacher should gradually create their own version.

- Start with a problem or situation that is meaningful to the students.
- Students solve the problem using their own strategy under teacher supervision.
- Multiple strategies are allowed.
- The strategies are discussed.

Introducing a situation in the classroom and allowing students to use their own strategies will lead to examples of the different stages in the development of a concept. When discussing student strategies, we should proceed from the most primitive to the most sophisticated. This promotes conceptual development following the logic of learning. This process also allows and encourages students to see other ways of solving the problem, sometimes more efficient than those they have used. The teacher may say, "Compare your strategy with Carla's. Which is more efficient?" This discussion also helps students become aware of their own strategies and how they relate to other methods. We emphasize the importance of dialogue in the classroom, both between students and teacher and among the students themselves. This helps students clarify doubts among themselves as well as to consolidate their knowledge by explaining it to others. In addition to the benefits of teamwork for learning mathematics, this process supports the development of attitudes and skills that encourage social interaction. Student interaction also stimulates oral expression, language development, and mathematical reasoning by the explaining of their strategies. In these exchanges, teacher intervention should target the encouragement of reflection, research, and proper language, instead of simply targeting the presentation of the solution. Through questions and answers, we should lead students to reflect and to realize the efficiency of certain strategies.

This method encourages both individual work and group work. Individual and group work are not incompatible. In fact, group activities may address different levels of comprehension, which allows students to go at their own pace. Including various levels of work in the same classroom is the most effective way of addressing individual differences. By letting each student work at his or her own level, we avoid segregation of students and we let more advanced students motivate the rest of the group.

In this book we offer tools to implement these strategies. However, we recognize that there is no single way to do this. Teachers should gradually construct, from the needs of their students and from their teaching experience, their own alternatives. For this, it is essential to observe and listen to the students, trying to understand the logic of their explanations and building upon them.

References

Bruner, J. 1996. The Culture of Education. Cambridge, Mass.: Harvard University Press.

Carey, S. 1985. Conceptual Change in Childhood. Cambridge, Mass.: MIT Press.

Davis, P.J. and Hersh, R. 1981. The Mathematical Experience. Boston, Mass.: Birkhäuser.

Freudenthal, H. 1991. Revisiting Mathematics Education. Dordrecht: Kluwer Academic Publishers.

Gravemeijer, K. 1994. Developing Realistic Mathematics Education. Dordrecht: Kluwer Academic Publishers.

Kuhn, T. 1977. The Essential Tension. Chicago: The University of Chicago Press.

Lakoff, G. 1987. Women, Fire and Dangerous Things: What Categories Reveal About the Mind. Chicago: University of Chicago Press.

Streefland, L. (ed.). 1991. Realistic Mathematics Education in Primary School. Utrecht, the Netherlands: Freudenthal Institute.

Treffers, A.E. 1987. Three Dimensional: A Model of Goal and Theory Description in Mathematics Education. Dordrecht: Kluwer Academic Publishers.

van den Heuvel-Panhuizen, M. 2003. "The didactical use of models in realistic mathematics education: An example from a longitudinal trajectory on percentage", Educational Studies in Mathematics, 54, 9–35.

Van Hiele, P.M. 1986. Structure and Insight: A Theory of Mathematics Education. Orlando, Fla.: Academic Press.

Wilder, R.L. 1973. Evolution of Mathematical Concepts. Cambridge, UK: The Open University Press.

CHAPTER 3

NUMBERING

3.1 Introduction

Numbers originated for the purpose of counting. Counting is one of the oldest mathematical tasks. Anthropologists have found evidence of some form of counting in all primitive cultures, although some of those forms are represented by only a few (numerical) words. Even some animals have the ability to distinguish different numbers of objects (Brannon *et al.*, 2001; Hauser *et al.*, 2000). The ability to count represents progress towards the development of number systems. According to the needs of each culture in using numbers, different number systems have been created (Wilder, 1968).

The first symbols used to express numbers were probably marks like I, II, III, IIII, to represent 1, 2, 3, and 4, respectively. This system is not efficient when we want to represent relatively large numbers. As cultures evolved and trade and agriculture began, using large numbers became necessary. This forced human beings to invent new ways of representing numbers.

Different cultures invented different number systems (Krusen, 1991). The number system we use today is the result of a long historical process. We give the name "number system" to the set of symbols used to represent numbers and the set of rules for combining these symbols to form other more complex symbols. The symbols we use to represent numbers are known as digits or numerals.

Just as in the historical process, learning to count is an important step in the formation of the concept of numbers in children (Fuson, 1988;

Fuson and Hall, 1983; Gelman and Gallistel, 1978; Resnick, 1983, 1989; Starkey and Cooper, 1983; Whalen *et al.*, 1999). However, unlike primitive cultures, the child, even before knowing how to count, runs into a language in which symbols and number words abound. In fact, the term "number concept" should be replaced by "number concepts," since the term "number" is used in different ways, such as:

- For reference (example: bus 14).
- For counting.
- For representing quantities.
- For measuring.
- As part of a system of conventions.

When children start school, they have probably encountered these different uses of numbers, but are unable to relate them. Their notions of number are closely linked to applications in the particular contexts where they have been learned. Teaching at the elementary level should seek to promote that students interrelate the different notions they have about numbers and develop a more mature notion that connects these different meanings of the concept.

3.2 Counting Numbers: The (Non-sensical) Number Sequence (Kindergarten to Six/Seven Years)

The first experience that children have with numbers is probably reciting the sequence of numbers. From a very small age in games and interactions with their elders, the sequence of numbers becomes a form of counting. Children first learn the sequence of numbers by the meaningless repetition of "one, two, three, …" It is like reciting a poem or singing a song. It is important to understand that at this stage children have not really learned how to count. However, it is beneficial to promote the learning of this sequence because it will help children to count later on.

3.2.1 *Sequence as an order in space*

Evidence from several studies indicates that upon entering school most children, based on their experience with the number sequence, have constructed a representation of numbers that can be characterized as a mental numerical sequence (the further to the right on the number line, the greater the number):

$$\longrightarrow \text{greater}$$
$$1 \rightarrow 2 \rightarrow 3 \rightarrow 4 \rightarrow 5 \rightarrow$$

Numbers correspond to places in a sequence, where each following number in the sequence is seen as greater than the previous ones. This representation is used both to count and to compare quantities.

Activities that Promote Learning the Number Sequence as an Order in Space

- Counting time in games and activities (e.g. while playing hide and seek)
- Counting when holding breath underwater
- Counting portions of food (e.g. counting raisins, candies, baby carrots).

In the process of repeating the number sequence, children are introduced to numerals. Again, even if students do not yet have a complete concept of numbers, learning the numerals from 1 to 10 will help them later on during the meaningful learning process.

3.2.2 *Sample activity*

Connect the dots. What do you see?

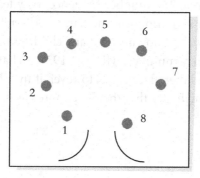

3.2.3 *Integrating metaphor: The number line*

The number line is one of the most commonly used metaphors in mathematics. As students are developing the idea of the sequence of numbers, the metaphor of a number line is introduced. Originally, counting is used as a step between beads and the number line:

1, 2, 3, 4, 5

Then we draw the beads with the number line so that students can associate the line with the number sequence:

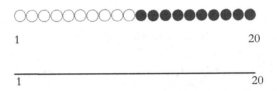

The line helps to see the order of numbers associated to distance in space: the farther apart, the greater the distance. This integrates the ideas of quantity and measurement. In Chapter 4, we will see how the number line also supports the development of the basic arithmetic operations.

3.3 Numbers as Quantity (Kindergarten to Six/Seven Years)

As children get to know the number sequence, they learn the concept of quantity. Yet, these two notions are not necessarily related in the child's mind. For decades, we've known that many children who "know" how to count do not conserve numbers (Piaget, 1952). That is, children may count five flowers and then five vases. However, if the objects are placed as shown below, they might say that there are more vases.

Hence, the following exchange is not surprising:

Teacher: How many flowers are there?
Child: Five.
Teacher: And vases?
Child: Five
Teacher: Are there more flowers or more vases?
Child: Vases.
Teacher: But there five of each?
Child: Yes, but there are more vases.

The student sees no contradiction here. In Piagetian terminology, this phenomenon has been described as the child "not conserving quantity." When children who make this mistake are asked about quantity they are fixing on other attributes. In the case of flowers, they are looking at the space they occupy. Thus, the first task we have when developing the idea of number as quantity (cardinality) is differentiating this attribute from others, like color, size, and spatial distribution. A basic concept needed to achieve this differentiation is the property of equivalence.

3.3.1 *Property of equivalence (1–1)*

▪ **Context of interest**
The need to compare quantities arises in many areas. For instance, are there more flowers than apples in the following picture? How do you know?

▪ **Allow students' informal methods and strategies**
As stated in the previous chapter, we start by allowing children to use their own strategies. For example, children might solve the problem above by qualitative comparison, choosing the one they perceive is a greater quantity. This strategy works when a set is clearly larger than the other. When we present two sets with similar amounts, this strategy fails because it is difficult to determine through qualitative comparison which one is larger. For instance, are there more spoons or cups in the following picture?

- **Encourage reflection and discussion about the proposed strategies**
  By analyzing students' strategies, we expect to reach the strategy of associating each spoon with one cup, that is, a 1–1 ratio.

- **More problems of a similar structure are introduced.**

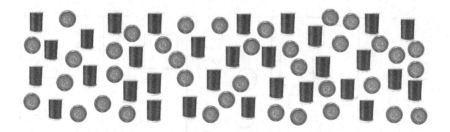

Note that so far we have not included numbers to compare quantities. In some of these problems, for example, strategies may arise such as counting by twos or by threes. In other words, when a 1–1 ratio is used, instead of associating each object with another object, we associate sets of two or three, as appropriate, with other sets of two or three, like in the following example.

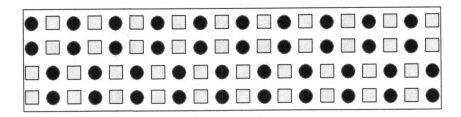

3.4 Associating the Number Sequence with the Idea of Quantity

As we work with the notions of number that students bring to school — number as sequence and number as quantity — we foster the analysis of properties that allow students, among other things, to begin associating these two notions into a more comprehensive concept of number. In fact, meaningful counting is just that, associating a sequence with a quantity.

Moreover, various dice games that require a peg to be moved are helpful in fostering the idea that numbers represent quantities. Hence, we introduce activities in which students need to associate a number with a set of objects.

Color the correct number

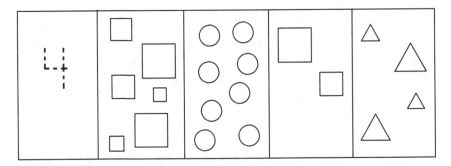

To be able to count correctly, in addition to making the association between the number sequence and the concept of quantity, it is necessary to master multiple processes, of which often we are not aware. When counting, it is necessary to be able to decide which elements belong to the set that we are counting. For example, children may have difficulty counting a particular set because the defining property is unclear. Hence, when evaluating certain elements, they cannot determine whether they belong to the set and, therefore, whether they should be included in the count.

In addition to deciding whether an element belongs to a set, it is necessary to keep track of the elements already counted. We often see children counting the elements of a set and they count the same element twice or more. Therefore, it is necessary to develop strategies that separate the counted elements from those not yet counted.

As students develop the ability to count, we introduce activities that require more efficient strategies, such as finding patterns.

Which side of the river has more flowers? How do you know?

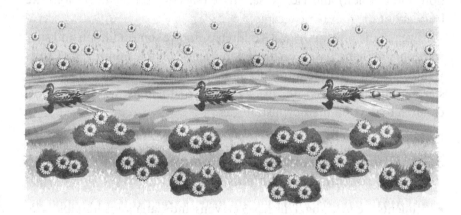

Another strategy that will support the development of multiplication later on is to group elements and count by twos, threes, fives, etc.

How much do they cost?

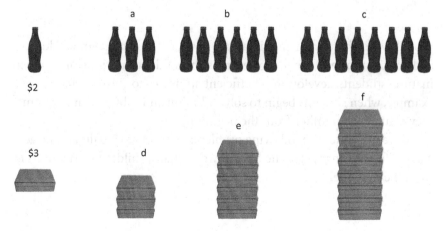

These last two exercises can be introduced to more advanced students, as a challenge, while the rest of the group acquires the ability to count.

3.5 Informal Arithmetic (Kindergarten to Eight Years)

As we stated in Chapter 1, teaching for understanding is not linear. Therefore, learning a concept supports, and is related to, the learning of other concepts. Thus, learning the number concept supports and enriches the learning of arithmetic operations (addition, subtraction, multiplication, and division) and vice versa. While reflecting on the operations, we learn about the number concept.

Even at the preschool level, children can solve some addition and subtraction problems based on their knowledge of the number sequence. Hence, it is important to introduce situations that require these operations starting in kindergarten. In this, as in other situations, we should allow children to develop their own strategies.

3.5.1 *Examples of contexts to introduce addition and subtraction problems*

- Students are painting. Luis has 5 crayons and Carla joins his table with 4 crayons. We ask: "If they share crayons, how many do they have between both of them?"
- Students are in their snack period. Robert has 6 cookies but Ana forgot her snack. We ask: "If Robert gives 2 cookies to Ana, how many are left for him?"

Likewise, there are many situations in which we can integrate addition and subtraction problems. However, as the students' notions of number mature, students develop more efficient strategies to solve problems. For example, when students begin to solve addition and subtraction problems, they count each number from the beginning.

If we introduce the following problem: "A child has 5 balls and he gets 4 as a gift. How many does he have now?" Initially, children start counting from the first one.

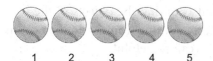 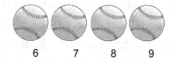

1   2   3   4   5   6   7   8   9

If we ask children to start counting from the number five onward, they may not be able to do it. When they grasp the concept that the last number on the count is the cardinality of the set, they can start counting from the last number of the first group.

The understanding of this property is not automatic. It requires that we develop activities to promote learning it.

3.5.2 *Examples of "Counting-from" Activities*

Introduce a problem that requires addition

For example, "Maria has 5 candies and her friend gives her 3 more. How many candies does Maria have now?"

We represent the problem as follows:

5   3

You will notice that many children start counting from the beginning. We can use the following metaphor to promote that the student starts counting from the number 5. We indicate that the first 5 elements of the sum will be placed inside a small box:

Students can knock five times as if counting the objects in the box. Gradually, they will realize they can start counting from 5 and then count the rest. Exercises like the following two can be helpful.

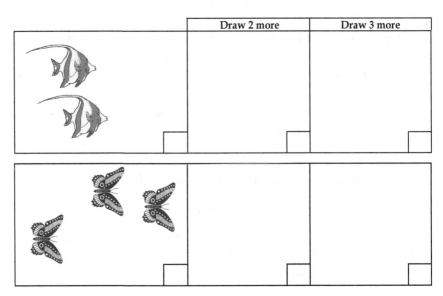

	Draw 2 more	Draw 3 more

Each number is one more than its predecessor (six/seven years)

In the process of working with whole numbers, one of the properties that should arise is that each number is one more than its predecessor. This seemingly simple property is not obvious for children. Make a ladder with Cuisenaire rods of the represented numbers, and then asking them to observe the regularity — that between each number, the difference is one block — may help students become aware of this property.

Understanding that each number is one more than its predecessor lets students transform addition and subtraction problems into simpler problems. For example, the sum 9 + 5 can be transformed into 9 + 1 + 4. In the next chapter we will discuss these strategies further.

The property of part–whole (from six years)

One of the major conceptual advances of children during their early grades is the ability to expand their interpretation of numbers to include it in terms of the part–whole relationship. In this way, children can think of a particular number as composed of two other numbers.

Many classroom situations lend themselves to the understanding of the part–whole relationship in numbers. For example, if a group of children is working at a table, you may ask: "How many children are at the table? How many boys? How many girls?"

Once several part–whole situations have been introduced, we incorporate exercises that support the development of this property.

(1) Show different ways of distributing the fish in the fish tanks.

---

(2) Show different ways to distribute the glasses on the tables. In how many ways can you do it?

---

(3) In how many ways can you distribute the flowers in the vases?

The traditional abacus is very useful for learning the concept of part–whole. In fact, we suggest the construction of an abacus for the early grades, as shown in the illustration below:

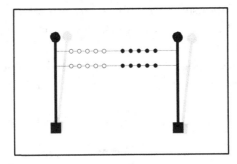

In this abacus we have 20 beads, arranged in two rows of 10 each. The rows in turn are subdivided into 5 white counts and 5 black counts. In the early grades, we must emphasize that students learn in depth the different relationships between numbers from 1 to 20. If children develop a solid base in understanding this, it will be easier for them to see how the patterns of relationships that exist between numbers are generalized into others. If they understand that $7 + 3 = 10$ then it will be easier to see why:

17 + 3 = 20,
27 + 3 = 30,
37 + 3 = 40, etc.

Here are other activities to promote the idea of part–whole.

Dominoes

In dominoes, students see numbers represented as combinations of two numbers.

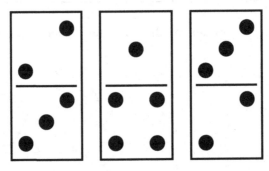

We could invent games that make use of this feature of domino tiles, like playing a game of "war" in which each child turns a domino and, when the same total comes up — for example, 2–5 and 3–4 — there is war. The next step is for children to turn a third domino each, and the one with the largest number on the third domino wins. If the same total comes up again, they may continue to draw tiles until one wins.

Although the idea of part–whole is introduced in first grade (six to seven years), it is important to practice these kinds of exercises for several grades, since, as we will discuss later, this is one of the basic ideas used in addition and also one of the fundamental ideas of our number system.

3.6 The Number System

Our number system is one of the most important inventions of human-kind. It has an impressive organization and structure. It is not found among the initial forms that humans used to express numbers, but rather shows maturity in mathematical development. Understanding this system is not easy; it requires maturity in understanding the concept of numbers. For example, it is necessary to become aware of several properties of numbers, as well as of the relationships between these properties. In this sense, we should not rush to introduce the notion of tens, hundreds, and so on until the student begins to understand the concepts on which these notions are constructed — that is, the idea of part–whole. In other words, the idea that each place represents a different type of object — groups of powers of ten. Although we discuss each of these concepts separately, the activities generally include several of these elements.

3.6.1 *The idea of part–whole*

Our system is based on the idea that one quantity can be expressed as the sum of others. Thus $245 = 200 + 40 + 5$. From first grade, we should develop the idea that a number can be decomposed into others. At the end of first grade, we start to introduce exercises in which the decomposition uses the number ten as one of its parts.

3.6.2 *Examples*

Situations that divide numbers into groups of ten

Introduce situations where dividing a quantity in groups of ten can help in the understanding of our system. These problems provide reflection on how we write numbers. For example:

> *A total of 76 students will participate in a school activity. The activity requires students to work in groups of 10. Each group of 10 students will have its own room. Once the students are divided into groups, if fewer than 10 of them remain for form one last group, then a group with fewer than 10 students will use the principal's office. How many rooms do we need to use? Do we have to use the principal's office? How many students will be in the principal's office?*

We should allow students to solve the problem with methods they understand, followed by a discussion of different strategies once the problem is solved. After the discussion, we should note the similarities between the result and the way the number is written.

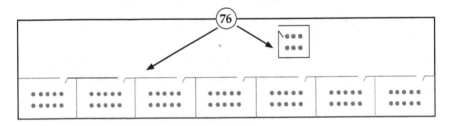

Seven rooms (with 10 students each) and 6 students in the principal's office

After introducing several of these problems, ask students: "Is the resemblance between the result and the way the number is written merely a coincidence?" This type of exercise helps students understand how numbers are represented. In fact, the child often does not naturally organize numbers in groups. This was made clear in an activity discussed in Chapter 2.

Counting by tens and hundreds

From second grade (seven to eight years), we work on activities that develop counting by tens as we saw in Chapter 2. In third grade (eight to nine years), while other decimal places are analyzed, problems are introduced to develop counting by hundreds.

3.6.3 *Each place represents a different type of object*

Another important feature of our system is that it is positional — that is, each place represents a different quantity, which is base 10 in our system. In Chapter 2, we showed how our monetary system helps in the understanding of this concept. Following are other representations that support this learning.

3.6.4 *Integrating metaphor*

There are certain contexts in which we can develop the idea of our number system that can serve as metaphors to help students understand the properties of our number system.

Cookie shop

The context of a cookie shop that sells loose cookies (units) in bags of ten (tens) and in boxes of ten bags (hundreds) is a very effective context to teach the positional value of our system (López and Velázquez, 2011). We begin by asking students to show how they would sell 15, 26, or 43 cookies (or any other amounts). The cookie shop metaphor also helps students develop the algorithms of addition and subtraction.

Once students have experienced these representations, we can introduce the Cuisenaire rods — where the model of our system is already given — by initially associating it with coins.

Note that we suggest a curriculum order that is different from the current one. In fact, we follow the order set out in the first chapter:

concrete → pre-formal → formal

We start by exploring numbers, observing their organization in groups of ten. It is not until students have this experience, as well as the understanding of the part–whole property, that we introduce the properties of our system. Initially we use metaphors (pre-formal) to analyze these properties: groups of straws, the monetary system, the abacus. We suggest not formalizing the idea of ones, tens, and hundreds until third grade (eight to nine years).

3.6.5 *Parallel forms to develop the idea of our number system*

As we argue at the beginning of the second part of this book, mathematical concepts are intertwined, which implies that sometimes when a topic is discussed, another topic is integrated or strengthened. The construction of various concepts interacts with the concept of our number system. We should take advantage of these situations to expand upon the understanding of our system. Let us see some examples.

The study of patterns in our number system

Recognizing numbers 11 to 20

Numbers from 10 on are formed by combining known digits. When we introduce them, we promote students observing the pattern we use to form them. This observation provides the basis for explaining our number system later on. For example, introducing numbers 10–19, we may ask: "What relation do you see between these numbers and the numbers from 1 to 9?"

Children will make different observations. Remember that we should not dismiss incorrect observations, but inquire as to why and how they propose them. Often children have a logic that we do not understand, but that is well reasoned from their perspective. Understanding this logic allows us to discover what children are thinking and thus help them develop a correct explanation.

From discussion with the children, the following pattern should arise:

1	2	3	4	5	6	7	8	9	10
↓	↓	↓	↓	↓	↓	↓	↓	↓	↓
11	12	13	14	15	16	17	18	19	20

Recognizing numbers 1–100

Recognizing that there are patterns when writing numbers helps children learn. Thus, once students know the numbers from 1 to 20, we ask: "What's next?" We continue with this exercise until we can make a chart representing numbers 1 to 100.

Once the chart is completed, we help students observe the following patterns:

- From left to right, numbers go: _1, _2, _3, ...
- From top to bottom, numbers go:

$$1\_$$
$$2\_$$
$$3\_$$
$$4\_$$
$$\vdots$$

Recognizing these patterns helps in number memorization and counting. It also helps students to formalize the idea of the structure of our system.

Arithmetic operations

Algorithms for solving the arithmetic operations are based on the properties of our number system. Therefore, developing these algorithms will also involve the application and strengthening of the understanding of the system. However, although logically the algorithms are based on the properties of the system, cognitively the operations are easier for students. As we discussed in the Chapter 2, **the logic of mathematics is not equal to the logic of learning**. This requires a change in the way algorithms are taught. In the chapters on operations, we will elaborate on the proposed change. Yet, we want to anticipate that reflecting upon these operations reinforces the learning of the properties of the number system. It is important to investigate this relationship.

Numbers as areas of exploration

As students become familiar with numbers, numbers become a context for exploration. For instance, we can and should encourage students to search for patterns among numbers. As part of student diversity, we will have in the classroom future mathematicians and mathematical scientists. It is important to challenge this group with problems and situations in which they start developing the elements of number theory. This is also an instance of diversifying the curriculum. For example, after studying our decimal system, we can present the following problems to students in fifth or sixth grade (10 to 12 years):

Imagine we went to a planet where the inhabitants had six fingers instead of ten. What do you think the number system would look like on this planet?

Discussing this situation leads to a system based on the number six. Thus we can see that all numbers can be expressed in any base greater than or equal to two.

Moreover, with this group of students we can also consider various number systems developed throughout history, such as the Roman system, and observe the advantages of having a positional system — that is, a system where each place represents a different quantity.

References

Brannon, E., Wusthhoff, C., Gallistel, C.R., *et al.* 2001. "Numerical subtraction in the pigeon: Evidence for a linear subjective number scale", Psychological Science, 12, 238–243.

Fuson, K.C. 1988. Children's Counting and Concepts of Number. New York: Springer-Verlag.

Fuson, K.C. and Hall, J.W. 1983. "The acquisition of early number word meaning: A conceptual analysis and review", in Ginsburg, H.P. (ed.), The Development of Mathematical Thinking. New York: Academic Press.

Gelman, R. and Gallistel, C.R. 1978. The Child's understanding of number. Cambridge, Mass.: Harvard University Press.

Hauser, M., Carey, S. and Hauser, L. 2000. "Spontaneous number representation in semi-free-ranging rhesus monkeys", Proceedings of the Royal Society of London: Biological Sciences, 267, 829–833.

Krusen, K. 1991. "A historical reconstruction of our number system" Arithmetic Teacher, 38(7), 46–48.

López, J.M and Velázquez, A. 2011. Contexts for column addition and subtraction. Teaching Children Mathematics, 540–548.

Piaget, J. 1952. The Child's Conception of Number. London: Routledge.

Resnick, L.B. 1983. "A developmental theory of number understanding", in Ginsburg, H.P. (ed.), The Development of Mathematical Thinking. New York: Academic Press.

Resnick, L.B. 1989. "Developing mathematical knowledge", American Psychologist, 44(2), 162–169.

Starkey, P. and Cooper, R.G., Jr. 1980. "Perception of numbers by human infants", Science, 210, 1033–1035.

Whalen, J., Gallistel, C.R. and Gelman, R. 1999. "Nonverbal counting in humans: The psycho-physics of number representation", Psychological Science, 10, 130–137.

Wilder, R. 1968. Evolution of Mathematical Concepts. New York: John Wiley and Sons.

CHAPTER 4

ADDITION AND SUBTRACTION

4.1 Introduction

Mathematical operations should arise from analyzing real situations. Starting in kindergarten, we should introduce children to problems that can be solved by addition or subtraction (Kamii, 2000; Carpenter and Moser, 1984). Initially, students will solve these problems by counting. In the previous chapter, we discussed how counting lends itself to solving addition and subtraction problems. We also analyzed how, as the their notion of numbers matures, students develop more efficient addition and subtraction strategies. Likewise, as students expand their development of addition and subtraction, they expand their notions of numbers and the numerical system.

In this chapter, we discuss how to support students' constructions of the concepts of addition and subtraction (Fuson, 1992). Given the close relationship between these operations, we should not separate their teaching. Instead, we should present situations that model addition and situations that model subtraction. Similarly, when working on addition exercises, we should include subtraction exercises. It is important to remember the close relationship between the notion of number and the basic operations; therefore, the trajectories of these two concepts are continuously intertwined during teaching.

4.2 Stages in Teaching These Operations

4.2.1 *The part–whole idea*

The idea that a number can be broken down into others is the basis of addition and subtraction. Thus, before introducing situations that portray these operations, students must have had some experience with problems that decompose numbers. For example, consider six bowling pins and two blank cards representing the decomposition.

In number decomposition problems, there are developmental stages. Initially, we introduce sets in which the number can be decomposed.

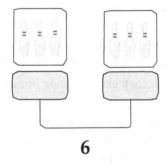

6

Then, one of the sets is replaced by a number.

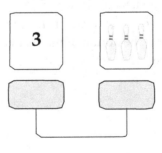

6

Then, we present the same framework, but the number is broken down into two other numbers.

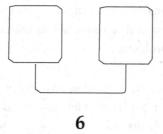

6

Finally, the number to be decomposed is presented, without reference to any particular context.

6

We also could use six-sided dice and ask students for ways of getting six dots.

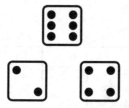

We can then substitute dice with numbers, letting students internalize that, for counting purposes, each face is equal to the number of dots. Moreover, dice can provide an easy transition from counting to addition and subtraction of numbers from one to six. Introducing 8-, 10-, or 20-sided dice can help us extend this example.

Notice how a concept we consider simple involves a process of construction. In this process, we give meaning to the sum combinations, since they are the way in which the number is broken down. In fact, repeating these problems is a way to automate the learning of sum combinations.

At a later stage, students will understand that the form to write numbers is based on their decomposition into groups of powers of ten. For example, they will understand that:

124 = one group of 100 + two groups of 10 + four 1s.

4.2.2 *Situations that model addition and subtraction*

Take a topic that is being discussed in class to introduce problems that require mathematical analysis that can be approached by addition or subtraction. For example, in a discussion about where students live, we can present the following problem:

> *There are students who live in the same area, but in different sectors, such as in a housing project. Some live in building A and others in building B. After counting those who live in building A and those who live in building B, we ask: "How many live in the project?"*

In addition to problems that arise from class context, we can use exercise sheets that present situations that require addition or subtraction, but without introducing the operation symbols. For example, how many more bags do we need to have eight of them?

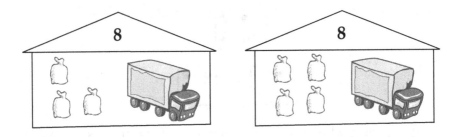

4.2.3 *Integrating models — metaphors for learning*

There are several integrating models that help students bridge their experiences with mathematical models. In the process of analyzing situations that show addition and subtraction, we present situations that may later become metaphors to serve as bridges between new situations and the operations of addition and subtraction. When students analyze these examples they might say, "Ah! This is like _____."

In the case of addition and subtraction, we can use the metaphor of buses picking up and dropping off passengers, or that of machines that perform an operation on numbers that are entered and then give a result.

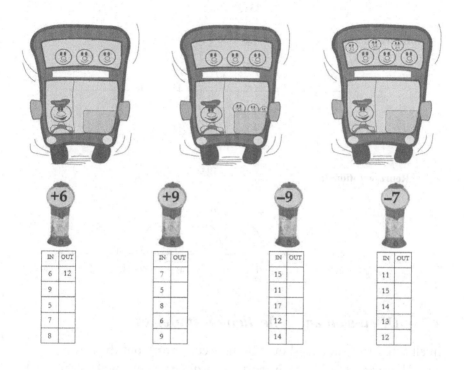

IN	OUT
6	12
9	
5	
7	
8	

IN	OUT
7	
5	
8	
6	
9	

IN	OUT
15	
11	
17	
12	
14	

IN	OUT
11	
15	
14	
13	
12	

Another representation that is very effective in supporting addition and subtraction is the empty number line (Steffe and Cobb, 1988). As noted in the previous chapter, the number line is one of the most useful representations in mathematics. In the case of operations, the line encompasses and supports relating the idea of numbers as a sequence with that of the numbers as a whole. For example, in the line we see the sequence of numbers, and we can observe that all 30s follow the 20s, and that these numbers come before the 40s. We also observe, for example, that 34 is equal to three groups of 10 and four units. Combining these two properties allows students to develop, from their original notion of numbers as a sequence, the idea of numbers as a combination of units, tens, and hundreds. Here are two possible representations of addition on the number line.

23+31

Representation 1:   **23 + 30 +1**

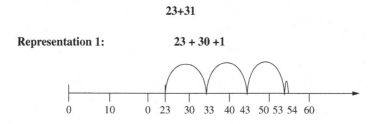

Representation 2:

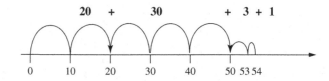

4.2.4 *Discussion and reflection on strategies*

In all activities, students should be allowed to work out their own strategies. However, it is just as important to discuss their methods through interaction between the teacher and other students. This fosters the development of their ability to describe and analyze their strategies. It allows students to become aware of other possible ways to solve problems more efficiently. For example, in the problem of children living in a housing project, let's say five children live in apartment A and six live in B. Let students develop their own solutions. We have seen that the following strategies arise in this problem:

— They draw five sticks and then six more. Then they count all of them.
— Starting from five, they count six more either with sticks or fingers.
— They use their hands as a counting instrument, along with number analysis, and say: "6 is one more than 5, so I have 5 and 5 plus 1. This is equal to 10, which are my two hands, plus 1, so it is 11."

Once students solve the problem, there should be a discussion of the different strategies that were offered. On the one hand, this allows us to see the analytical level of each child. On the other, we foster group learning. For example, in the previous problem, the same children can become aware that decomposing the numbers is a good strategy for addition.

The art of asking questions is very important in this process. Questions should not be aimed at eliciting definite answers, but rather at promoting reflection. The discussion should stimulate the use of mathematical language by students, making them aware that there are multiple ways of solving a problem.

4.2.5 *Symbolism of addition and subtraction*

Symbolism in addition and subtraction, as well as in other arithmetic operations, should be gradually constructed. We start with a qualitative analysis of the situations. Once we have worked with several cases, including integrating models, we use them to introduce representations accompanied by numbers and an intermediate symbol, like an arrow.

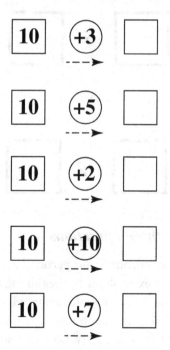

After the student has worked with intermediate symbols, we introduce numerical problems with the addition symbol, accompanied by representations.

Finally, we introduce problems where only numbers and the addition symbol appear.

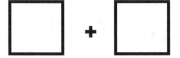

4.2.6 *Properties of these operations*

From the early grades, we encounter properties that are very useful in addition and subtraction. For example, at the beginning of this section, we discussed the idea that a number can be decomposed in different ways, which is quite useful for addition and subtraction. In the previous chapter, we mentioned the counting-from property. Let us see some other properties.

Using ten and five as references for addition and subtraction

From first grade (six to seven years), we promote the use of five and ten as reference numbers for addition and subtraction. When adding two numbers, we break them down by using five and ten as a basis. For example:

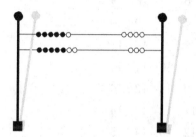 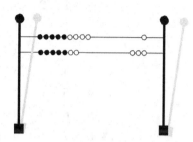

4.3 Working with Numbers

Sticking to problems in which students have to interpret situations, we gradually broaden the study of operations directly with numbers. An important task in this regard is the process of automating the basic data of these operations, which will later help students in the learning of other operations. In particular, students should learn about automating basic combinations.

4.3.1 *Basic combinations*

Number decomposition problems serve as a basis for learning basic combinations. The process of automating these combinations can by simplified if students discover various patterns.

Playing with five and ten

If we associate the operations with the properties of decomposing numbers (the part–whole idea) and, in this relation, we take as point of reference the five and the ten, we can foster the automating of basic combinations. Here, are some examples:

$$7 + 4 = 7 + 3 + 1 = 10 + 1 = 11$$
$$5 + 7 = 5 + 5 + 2 = 10 + 2 = 12$$

Pattern analysis

Patterns also reinforce the automating of combinations. For example, we can introduce an addition table to study the patterns involved.

+	1	2	3	4	5	6	7	8	9
1	2	3	4	5	6	7	8	9	10
2	3	4	5	6	7	8	9	10	11
3	4	5	6	7	8	9	10	11	12
4	5	6	7	8	9	10	11	12	13
5	6	7	8	9	10	11	12	13	14
6	7	8	9	10	11	12	13	14	15
7	8	9	10	11	12	13	14	15	16
8	9	10	11	12	13	14	15	16	17
9	10	11	12	13	14	15	16	17	18

The study of patterns should take place in discussion with students. Once we discover patterns, we should explain why they behave the way they do. Among the patterns that should arise are:

Each number is 1 more than the previous one.

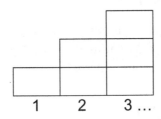

From this property, it should be evident that each time we add a particular number to two other consecutive numbers, the difference of their sums is one.

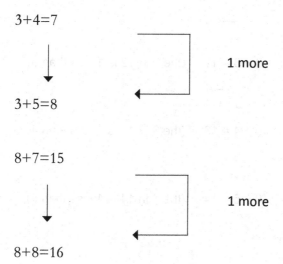

$$3+4=7$$

$$3+5=8$$

1 more

$$8+7=15$$

$$8+8=16$$

1 more

Addition is commutative

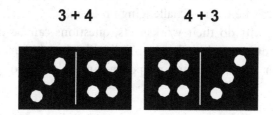

3 + 4   **4 + 3**

Hence, we only need to learn half of the combinations.

Property of nine

Nine is one less than ten.

Thus, when adding nine to another number, the result is one less than when adding ten.

9 + 3 = 12   (the 2 in 12 is 1 less than 3)

9 + 4 = 13   (the 3 in 13 is 1 less than 4)

9 + 5 = 14   (the 4 in 14 is 1 less than 5)

Memorization

Students should learn basic combinations from memory. The study of properties and patterns supports memorization; practice also helps. We can make this process flow more smoothly if we use games or interesting worksheets. Annex 1 includes several sheets of various difficulty levels. The harder sheets can be used as challenging problems.

While students do their worksheets, questions can be integrated to lead the students towards reflection. This reflection, in turn, allows students to better understand the properties and relations of the number system.

4.4 Situations that Model Addition and Subtraction

Addition and subtraction exemplify different situations. As the students' understanding of these operations grows, we start to integrate situations that model operations requiring more maturity.

Here is a table that presents different situations modeled by these two operations.

Category	Addition Problems	Subtraction Problems
Joining and separating	Luis has *a* cents. If his father gives him *b* more cents, how many cents does Luis have now?	Carmen had *a* lollipops and gave *b* to Heidi. How many lollipops does Carmen have now?
Part–whole	A group of children was at the playground. There were *a* girls and *b* boys. How many children were there in all?	There were *c* children at the playground. There were *a* boys and the rest were girls. How many girls were at the playground?
Comparison	Carlos has *a* toy cars. Jamie has *b* more than Carlos. How many toys cars does Jamie have?	During the field day, Mark won *a* prizes and his sister Laura won *c* prizes. How many more prizes than Laura did Mark win?
Equaling	Mary collected *a* flowers and Amy collected *c* flowers. What does Mary have to do to have the same number of flowers as Amy? (Amy picked up more flowers.)	Anna has *a* balls. Ralph has *c* balls. What does Ralph have to do to have the same number of balls than Anna? (Ralph has more balls.)

These problems present situations that offer different levels of difficulty. Studies show that situations about joining and separating are the easiest for children, whereas those comparing subtraction are the hardest. That is why we should introduce them in different school levels, or at least in different semesters. However, by third grade (eight to nine years) students should be familiar with different situations representing addition and subtraction.

A potential research project could examine whether learning is promoted when students see the structure of joining and separating in these diverse types of problems.

4.4.1 *Problem development by students*

While we present problems for students to solve, we should integrate activities in which the process is reversed; that is, formulate problems of mathematical situations. For example, consider the following operation:

$$
\begin{array}{r}
3 \\
+5 \\
\hline
8
\end{array}
$$

We can ask students to create a problem for which that operation is its mathematical representation. While working on different problems, we can ask students to create situations that are represented by a particular mathematical operation. This skill is essential for mathematical modeling and for learning to see the world through mathematical eyes.

4.5 Mental Arithmetic

Mental arithmetic, as the term suggests, is the ability to work arithmetical calculations mentally. For this, we use number properties that facilitate calculation, like decomposition, associativity, commutativity, and operations with multiples of ten.

For example, when adding 38 + 27, we can use number decomposition to convert one of the numbers into a multiple of ten, thus facilitating the calculation.

$$38 + 27 = 38 + (2 + 25) = (38 + 2) + 25 = 40 + 25 = 65.$$

Another way of thinking about it would be:

$$38 + 27 = (20 + 30) + (8 + 7) = 50 + 15 = 65.$$

Nowadays, when calculators can easily do the operations required by difficult calculations, the competence that we must develop in students is the ability to decide mentally if the calculation given by the calculator is reasonable. In this process, mental arithmetic is very important.

Like all competences and concepts, development of mental arithmetic is a gradual process. While we work at various stages in the development of operations, we integrate exercises that require mental arithmetic. For example:

- When working with basic combinations, we foster mental arithmetic as follows:

  $7 + 8 = 7 + (3 + 5) = 10 + 5 = 15$.

  Once we know that $7 + 8 = 15$, we can see that

  $8 + 8 = 8 + (7 + 1) = 15 + 1 = 16$.

- When discussing addition in columns, for example

$$
\begin{array}{r}
24 \\
42 \\
+36 \\
\hline
\end{array}
$$

We think $24 + 36 = 60$ and $60 + 42 = 102$.

4.6 Addition and Subtraction Algorithms

Once students understand what the addition and subtraction operations represent, we start teaching algorithms to carry out these operations. This is often done by introducing a set of rules without understanding. Students see these operations as a set of rules to follow. When analyzing some errors made by students with these algorithms (Quintero, 1986), we see that they start from an alternate rule developed by the students, and which they consistently follow. Here are some of the different types of errors we have found. Try to figure out what rule the student is following.

Error 1

5	34	28	43	56
+7	+21	+53	+26	+24
12	10	18	15	17

Error 2

5	34	28	43	56
+7	+21	+53	+26	+24
12	55	711	69	710

Error 3

5	34	28	72	63	73 ·
+7	+21	+53	+87	+57	+72
12	55	711	510	111	46

Error 4

5	34	28	54	38
+7	+21	+ 5	+26	+26
12	55	51	71	91

In Error 1, students add all the digits. The fact that each digit is in a specific position bears no importance to them.

In Error 2, students add each column without regrouping.

In Error 3, they add from left to right, following the rules given to add from right to left. That is, they carry from the tens to the ones.

In Error 4, when adding and getting a number greater than 10, they write the ten and carry the ones.

When analyzing these errors, we see that students have replaced one rule with another. When they learn rules without understanding, it is only natural for these errors to occur. Since these rules lack meaning for them, they will focus more on some of the details, and their rule becomes whatever they remember (Kamii and Dominick, 1998; Kouba *et al.*, 1997). It has been stated that, to avoid making the addition and subtraction algorithms look like a series of meaningless rules, it is necessary for students to participate in their construction (Kamii *et al.*, 1993). In this construction, it is necessary to follow the logic of learning.

Beishuizen and Anghileri (1998) state that students understand the addition model first, which arises from the sequence of the number line, and then the algorithm, which arises from the properties of our system. In fact, working with addition in the number line helps students explore the elements and properties integrated in the algorithm and, gradually, it helps them understand the algorithm itself. To achieve this, the process needs to be one of searching and reflection. Thus we gradually introduce the elements needed for students to construct the algorithm. They are discussed below.

4.6.1 *Jumping by tens*

The idea that ten is a fundamental element of our system is not automatic for children (Cobb and Wheatley, 1988). It is important to start developing this idea. Once students have practiced addition up to 20, for example, 7 + 8, 4 + 9, or 3 + 5, we start introducing problems that jump by tens on the number line. For example:

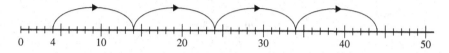

We also introduce jumping backwards, which lays the groundwork for the subtraction algorithm.

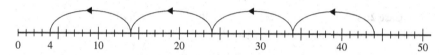

Likewise, we introduce problems in which both addends have no ones, for instance:

$$10 + 30 = \qquad 20 + 30 = \qquad 40 + 50 =$$

In subtraction:

$$30 - 10 = \qquad 50 - 20 = \qquad 60 - 40 =$$

Students can do these operations on the number line.

4.6.2 *Addition and subtraction with a two-digit number and a one-digit number*

Once students have some experience with the problems described above, we start working with problems in which one of the numbers has two digits and the other just one digit.

$$25 + 4 = \qquad 32 + 7 = \qquad 56 + 6 =$$
$$38 - 4 = \qquad 45 - 3 = \qquad 24 - 6 =$$

4.6.3 *Addition of two two-digit numbers*

Consider a problem like 23 + 31. We promote that students solve this addition problem by their own methods (Carpenter *et al.*, 1998; Fuson *et al.*, 1997). Here are two possible strategies to carry out this sum on the number line:

Case 1:

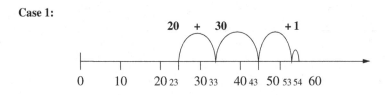

Case 2:

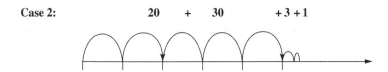

We allow students to solve addition and subtraction problems using the number line and ask them to explain the reasoning of their process. Thus, the elements that constitute the algorithm arise.

Another metaphor we can use to teach addition and subtraction algorithms is the cookie shop (López and Velázquez, 2011). As we explained in the previous chapter, the shop sells loose cookies in bags of 10 and also in boxes of 10 bags each. We introduce the situation of a girl who asks for 25 cookies and then asks for 37 cookies. If her requests are joined in one order, how many bags of 10 will she get? How many loose cookies will she get?

25 cookies 37 cookies

joining them

or

62 cookies

Once students have some experience with the number line and the cookie metaphor, around third grade (eight to nine years), when we have already started discussing ones, tens, and hundreds, we introduce the addition algorithm as a strategy that a child used in another classroom.

$$\begin{array}{r} 24 \\ +32 \\ \hline 56 \end{array}$$

We ask the children: "Is this process correct? How do you explain it? How does the strategy of the number line relate to the strategy of the cookies?"

The process of developing these algorithms can help students in understanding the properties of our system. The study of our system, in turn, explains the algorithm. We must analyze at what moment we should use each one to explain the other.

Research Questions

1. Does the process of students constructing the algorithms help them to understand our number system?
2. When should the number system be introduced as a justification for the algorithms?

Our hypothesis is that the relation between these two processes should be taught in spiral form. Initially, students have a better notion of addition and subtraction than of our number system. At that moment, trying to develop an algorithm to carry out these operations can help widen their notion of our decimal system. Once we introduce the algorithm used in mathematics, we can justify it by using the decimal system. We represent this relation in the following way:

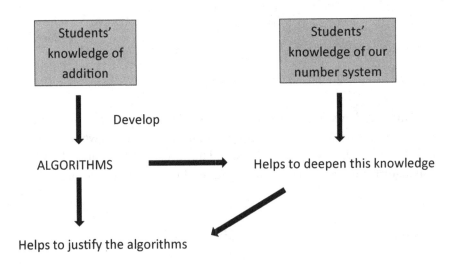

Once students have developed the addition or subtraction algorithm, we ask them to justify it on the basis of the properties of our system. In this process, we can use concrete materials like the Cuisenaire rods. Below is a presentation of a set of activities geared towards understanding the addition and subtraction algorithms using Cuisenaire rods to show the properties of our system.

Addition

Addition without carrying

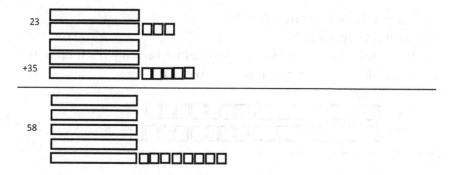

Students show the answer with Cuisenaire rods and, after several examples, observe that in addition we can add the corresponding units.

Addition with carrying

$$17$$
$$+18$$

Initially, children may write:

$$17$$
$$+18$$
$$115$$

We may say: "Do you think 17 + 18 is greater than 100? Try with Cuisenaire rods."

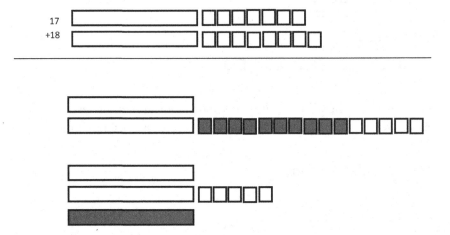

"Count. How many are there? 35.

What is happening?"

We explain that in the place of the ones we can only reach up to the number 9, thus, we must exchange ones for tens:

After several examples with the rods, we can go on to explain the carrying algorithm. In these examples, we should include cases with three numbers that carry more than 1. Once students understand the algorithm, practice exercises are given.

4.6.4 *Subtraction*

As we did with addition, we must allow students to develop their own algorithm. Once students have worked with their algorithm, we hold a group discussion in which all the children present their methods and the reasons for their use. For example, a study from the Netherlands (Streefland, 1991) found that children developed the following algorithms.

Algorithm: Addition

Some students solve subtraction problems with addition. For instance, when doing

$$64$$
$$-37$$

they think: 37 + **3** is 40 + **24** is 64. Hence,

$$64$$
$$-37$$
$$27$$

Algorithm: subtract and then regroup

$$346$$
$$-183$$
$$+2 - 4 + 3$$

$$200 - 40 + 3 = 163$$

In this case, we compare the numbers in each column. The intermediate result is the difference between the two numbers in the same column. If the minuend is greater, then the result is positive as there is an amount left over; if it is smaller, the result is negative since a certain amount is missing. Then, the intermediate results are added according to the positional value of the column. (Note: this algorithm requires a profound understanding of the positional number system and an intuitive knowledge of how to work with negative numbers; those skills are not trivial.)

Once students have presented their algorithms, we discuss them. If the standard method of subtraction by units does not arise, we introduce it as a method that was presented by a child in another class.

Find: 58 – 24
Algorithm: 50 – 20 = 30 and 8 – 4 = 4, so
58 – 24 = 34.

After explaining this method, we use Cuisenaire rods to justify it.

Subtraction without borrowing

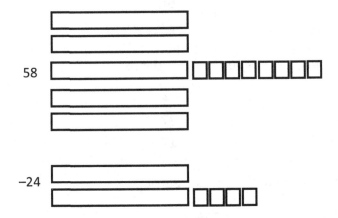

The child represents both numbers with Cuisenaire rods. We explain that in subtraction, we must take away from the first quantity an amount equivalent to the second one.

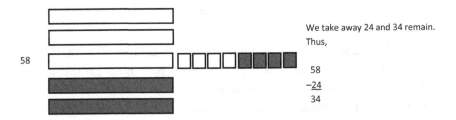

We take away 24 and 34 remain. Thus,

58
−24
34

After several examples, children will see that, in subtraction, we subtract the digits that correspond to the same unit. Some children will need to work out more examples to see this pattern and discover the rule. For example, after children have worked out several problems, the teacher may ask: "Do you see a pattern?" If they do not, we may suggest the rule and allow them to verify it with other examples. However, this is a last option since we would be depriving the child of the experience and thrill of discovery.

Subtraction with borrowing

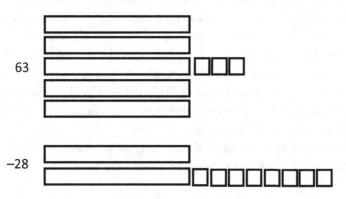

When we introduce this problem, students will try to solve it in the same way as the previous ones. A child might even say that it cannot be done. We discuss the problem and suggest that we can exchange a ten for ten ones.

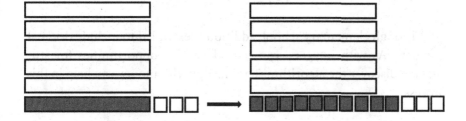

Now, we apply the previous method and find 35 for the answer.

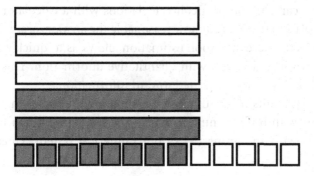

After several examples, we explain the algorithm referencing the Cuisenaire rods. These activities take more time than simply introducing the algorithms directly. In the long run, however, this time is recovered as students learn these algorithms with meaning and without the need to put students through endless remedial sessions. In fact, student errors show that learning these operations as a set of meaningless rules makes students confuse one rule with another, leading to incorrect algorithms.

Nevertheless, we must realize that the introduction of concrete models can turn into a series of meaningless games. Students can find rules and work with these models mechanically. This is why the teacher's role is extremely important as an agent who promotes reflection on these models with the numbers they represent.

Once students understand these rules, we present exercises for practice. These exercises should always promote reflection.

$$8 - 3 = \qquad 9 - 3 =$$
$$8 - 4 = \qquad 9 - 4 =$$
$$8 - 5 = \qquad 9 - 5 =$$
$$8 - 6 = \qquad 9 - 6 =$$
$$8 - 7 = \qquad 9 - 7 =$$

Through elementary school, addition is extended to include decimal numbers. As with integers, you should provide opportunities for students to discover the algorithm. The analogy with money can help in this process.

4.7 Estimation

The ease of carrying out difficult calculations with a calculator or computer forces us to think about how reasonable the answers are. To this end, estimating is the necessary tool. Estimation allows us to quickly grasp the reality shown in a situation with quantitative information. This is a very important skill in a society geared toward information.

Although adults might think that estimating is simpler than making calculations with difficult numbers, it is not easy for children because it requires knowing the properties of our numerical system. Thus, as with other competences, the development of estimating should be a gradual process (van den Heuvel-Panhuizen, 2001).

4.7.1 *The path to estimating*

(1) *Mental arithmetic*
Estimating requires mental arithmetic, which starts in first grade (six to seven years) when children are learning the basic combinations of addition. As students advance in their calculation expertise, they also develop mental arithmetic with those calculations.

(2) *Rounding*
Around third grade, tasks are introduced to develop the concept of rounding. As with other concepts, we start by introducing situations that make sense to students as a way to develop the intuitive idea of rounding. For example:

— You are interested in buying a toy that costs $8.40. Your mom only has $10 bills. How many bills do you need to take?
— There is a gas station on the road every 10 miles. Mario suddenly realizes that he has almost run out of gas. He has covered 23 miles. To which gas station should he go?

Once we have worked on several activities in context, we develop the rounding rule. This rule is based on the basic principle of determining the number (ten, hundred, thousand, etc.) that is closer to a given number. When applying the rounding rule, it is important to identify which unit is more reasonable for rounding. The curriculum should offer an opportunity to work on problems in which part of the task is to decide by which unit to round.

(3) *Estimating*
Estimation tasks begin in fourth grade (nine to ten years). There are two contexts that we use for estimating. In one of them, we have exact numbers and we can make exact calculations, though it is not necessary. For example, I am going to a party and they have asked me to bring four soda

bottles. I see the bottles on sale at $0.98. How many dollars do I need to take? I don't need to make an exact calculation to know that I need $4.00 or that I can take four $1.00 bills.

Another situation is that in which we have exact numbers for exact calculations, but we make an estimate. This happens when we want to verify how reasonable a result is, whether offered by someone else or by a calculator. For example, I am going to pay for four articles that cost $4.89, $5.10, $10.03, and $12.68. The cashier tells me that the total is $52.70. Is that a reasonable result? A quick estimate reveals that the sum is incorrect. If there are two articles ($4.89 and $5.10) whose sum is approximately $10, and the other two articles are close to $10, the total must be close to $30.

When solving problems with difficult calculations for which a calculator is used, we should always ask students to evaluate how reasonable the result is by estimating the answer.

Another context in which estimating is used is when we present estimated numbers instead of exact numbers. In that case, we also estimate the result. Unlike the previous cases, this one offers no possibility to make exact calculations. For example, in a political party rally we calculate crowds by estimating how many people fit, let us say, in a square meter and how many square meters there are in the rally area.

4.7.2 *Uses of estimation*

(1) *To confirm how reasonable a result is*
When using a hand calculator, it is always a good idea to confirm whether the result is reasonable to avoid errors. We use estimation for this purpose.

(2) *To quickly grasp the reality shown in a situation with quantitative information*
When we hear news or conversations that include quantitative data, it is important to estimate results quickly in relation to the information. For example, we receive several offers for a car rental. Depending on the use of the car, we can quickly estimate which offer is more suitable (see Annex 2, Estimation).

(3) *For a didactic purpose*
In general, when estimating and calculating, number connections become more visible, allowing for a broader knowledge of the structure of our number system. Because of the need to know the structure of the system, estimation can be used as a way to assess student comprehension of the material discussed. Thus, it is an assessment tool.

4.8 Hand Calculators

Calculators have several functions. Firstly, they can be used for support in calculations that take a long time. Secondly, they can also be a didactic tool that supports the teacher in creating situations that foster reflection on the structure of our number system. Finally, they are useful for solving problems and in the identification of patterns. Let us illustrate each one of these functions.

Just as with concepts and processes, calculator use should follow a gradual process and spiral development. When we introduce a calculator, we explain its functions in relation to the operations that we will use — originally addition and subtraction. We work with several functions of the calculator for addition and subtraction. We later repeat this process with multiplication and division.

4.8.1 *Support for calculations that take a long time*

Calculators help simplify numerical calculations that are time-consuming. However, the calculator should not be a substitute that replaces the learning of combinations and algorithms. Learning combinations and algorithms allows students to acknowledge the structure of the number system and, thus, carry out other tasks like mental arithmetic and estimation. Once students know the combinations and understand the algorithms, calculators can be used to simplify operations with numbers that are too large or for operations involving many numbers (see Annex 2, Activities 3, 4, 5, and 6).

4.8.2 *Didactic tool*

Calculators can support teachers in several tasks. For example, teachers can ask students to use a calculator to check their work or students can check their assignments with help from the calculator. This task is interesting for students, and it frees up time for teachers that would otherwise be spent checking the assignments of each student. Calculators also allow for the creation of situations that foster reflection on the structure of our number system (see Annex 2, Activities 1, 2, 9, 10, 11, and 12).

4.8.3 *Support in solving problems and identifying patterns*

Calculators can support students in identifying patterns and in problem solving (see Annex 2, Activities 7 and 8).

4.8.4 *Sample activities*

Throughout the school year, students will often use the calculator for some of these purposes. In Annex 2, we have included a work manual produced by the CEFAM (Centro de Estudio para Facilitar el Aprendizaje de la Matemática) project, developed at the Lab School of the University of Puerto Rico from 1987–1990, dedicated to presenting problems for calculator use. Activities 3, 4, 5, and 6 present problems that include calculations that take a considerable amount of time, for which the calculator serves as a tool. Activities 1, 2, 7, 8, 9, 10, 11, and 12 present problems that foster looking for patterns and relations that include addition and subtraction. Many of these exercises can be used as challenging problems.

References

Beishuizen, M. and Anghileri, J. 1998. "Which mental strategies in the early number curriculum? A comparison of British ideas and Dutch views", British Educational Research Journal, 24(5), 519–538.

Carpenter, T.P., Franke, M.L., Jacobs, V.R., *et al.* 1998. "A longitudinal study of invention and understanding in children's multidigit addition and subtraction", Journal for Research in Mathematics Education, 29, 3–20.

Carpenter, T.P. and Moser, J.M. 1984. "The acquisition of addition and subtraction concepts in grades one through three" Journal for Research in Mathematics Education, 15, 179–202.

Cobb, P. and Wheatley, G. 1988. "Children's initial understandings of ten", Focus on Learning Problems in Mathematics, 10(3), 1–28.

Fuson, K.C. 1992. "Research on learning and teaching addition and subtraction of whole numbers", in Leinhardt, G., Putnam, R. and Hattrup, R.A. (eds.), Analysis of Arithmetic for Mathematics Teaching. Hillsdale, N.J.: Lawrence Erlbaum Associates, pp. 53–188.

Fuson, K.C., Wearne, D., Hiebert, J.C., et al. 1997. "Children's conceptual structures for multidigit numbers and methods of multidigit addition and subtraction", Journal for Research in Mathematics Education, 28, 130–162.

Kamii, C. and Dominick, A. 1998. "The harmful effects of algorithms in grades 1–4", in Morrow (ed.), The Teaching and Learning of Algorithms in School Mathematics, 1998 Yearbook of the National Council of Teachers of Mathematics. Reston, Va.: National Council of Teachers of Mathematics, pp. 130–140.

Kamii, C., Lewis, B.A. and Livingston, S.J. 1993. "Primary arithmetic: Children inventing their own procedures", Arithmetic Teacher, 41, 200–203.

Kamii, C.K. 2000. Young Children Reinvent Arithmetic: Implications of Piaget's Theory, 2nd edition. New York: Teachers College Press.

Kouba, V.L., Zawojewski, J.S. and Strutchens, M.E. 1997. "What do students know about numbers and operations?" in Kenney, P.A. and Silver, E.A. (eds.), Results from the Sixth Mathematics Assessment of the National Assessment of Educational Progress. Reston, Va.: National Council of Teachers of Mathematics, pp. 87–140.

López, J.M and Velázquez, A. 2011. Contexts for column addition and subtraction. Teaching Children Mathematics, 540–548.

Quintero, A.H. 1986. ¿Qué me pasa con la matemática? San Juan: Editorial de la Universidad de Puerto Rico.

Steffe, L.P. and Cobb, P. 1988. Construction of Arithmetical Meanings and Strategies. New York: Springer-Verlag.

Streefland, L. (ed.). 1991. Realistic Mathematics Education in Primary School. Utrecht, the Netherlands: Freudenthal Institute.

van den Heuvel-Panhuizen, M. (ed.). 2001. Los niños aprenden matemática. Utrecht, the Netherlands: Freudenthal Institute. (Spanish translation by Project CRAIM, Universidad de Puerto Rico.)

ANNEX 1

ACTIVITY AND CHALLENGE SHEETS

Sheet 1

Complete each addition and subtraction puzzle.

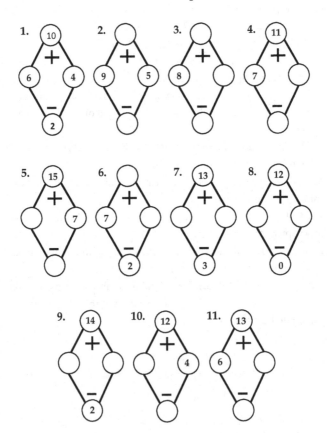

Sheet 2

Use numbers 1 through 9 to fill in the squares. The sum in each row and column should be 15.

1.

8	1	
3		
	9	2

2.

	3	
9	5	
2		

3.

7		5
2		
	8	

4.

2		
	5	3
		8

5.

	7	2
	5	
8		

6.

4		8
2		6

7.

6	1	
7		
	9	

8.

8		
	5	7
4		

9.

2		
		5
6	1	

10.

		6
3		
4		2

11.

6		2
8		4

12.

		8
9		1
	7	

Sheet 3

Add down and across.

1. (+) →

(+)	3	6	9
	2	1	3
	5	7	12

2. (+) →

(+)	5	1	
	3	4	

3. (+) →

(+)	8		9
	2		8

4. (+) →

(+)		2	6
	8		12

5. (+) →

(+)			8
	2	1	
	5		

6. (+) →

(+)		3	
		6	8
			16

7. (+) →

(+)	5		8
		6	15

8. (+) →

(+)			10
		1	
	9		18

9. (+) →

(+)	2		
			9
	6		14

Sheet 4

What path leads to the final number? Use a calculator to add or subtract. Draw a line along the correct path.

1.

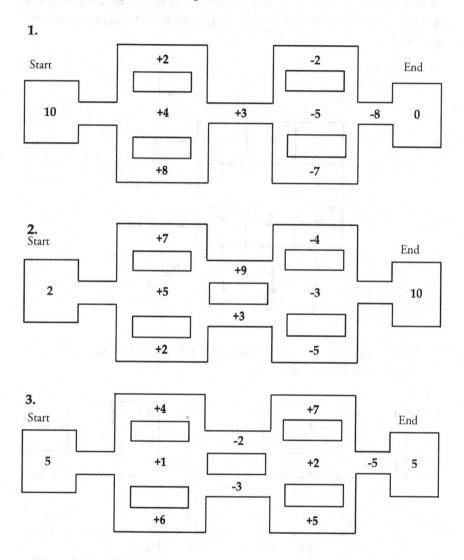

Sheet 5

Magic squares

The numbers in the squares below make a **magic square** because the sums of all columns, rows, and diagonals are the same. The magic total of this square is 18.

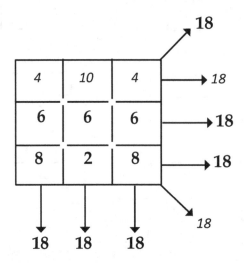

Make magic squares by filling in the empty squares with numbers. HINT: First find the magic total.

1.

9	2	
	7	6
4		5

2.

4		
2	3	
3		2

3.

		6
	4	
2		3

4.

6		
	5	3
2		4

Sheet 6

Use the numbers from the first rectangle to complete the square with magic sum 34.

1	2	3	5	7
9	10	12	15	16

		14	4
	6		
8		11	
13			

ANNEX 2

ACTIVITIES WITH A CALCULATOR

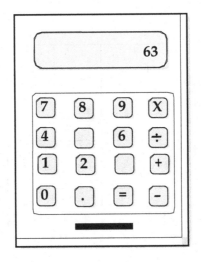

Activity 1

Instructions: Enter each of the following numbers in a calculator. Then, subtracting just one number, make the number in the box a 0.

2741	[7]		743021	[3]
80692	[9]		743021	[4]
348967	[4]		743021	[7]

Activity 2

Concept: Relation between positional places.

Enter each number in the calculator. Multiply the number by 10 several times.

8
9
18
15
16

What happens to each digit? Why do the numbers you've entered keep moving to the left?

Activity 3

Personalized problems.

Check your pulse for three to four minutes.

1. Find your average number of heartbeats per minute.
2. Find your number of heartbeats in one hour.
3. Find your number of heartbeats in a day.
4. Find your number of heartbeats from birth to your 40th birthday.

Activity 4

Personalized problems.

1. Write down the number breaths you take in three or four minutes.
2. How many breaths do you take in a day?
3. How many breaths have you taken in your life?
4. How many breaths will you take from now until Christmas (December 25th)?

Activity 5

Personalized problems.

Measure your height in centimeters. Use your calculator you help you count.

1. How many times does your height fit in a kilometer?
2. How many times your height is the height of your school building?

Activity 6

Personalized problems.

1. How old are you, in hours?
2. How old are you, in minutes?

Activity 7

1. Fill in the blanks. The odd numbers are:

    1, 3, ____, 7, ____, ____, 13, ____, ____, ____, ____, …

2. These are some examples of even numbers expressed as the sum of four odd numbers.

$$1 + 1 + 1 + 3 = 6,$$
$$1 + 3 + 3 + 3 = 10,$$
$$3 + 3 + 3 + 5 = 14.$$

- Use your calculator to find all possible ways of expressing the following even numbers as the sum of four odd numbers. Write down all the answers.
  a. 4.
  b. 6.
  c. 8.
  d. 10.
  e. 12.
  f. 14.

- Repeat b through f, but this time try to express each even number as the sum of six odd numbers.

Activity 8

Four magic digits.

1. Choose four digits, like 2, 3, 4, and 5, and follow the steps below:
   a. Make the largest number you can   5432
      Make the smallest number you can   − 2345
      Subtract the smaller from the greater   3087
   b. Take the digits from the difference (3, 0, 8, 7) and repeat the previous steps.
      Largest possible number   8730
      Smallest possible number   − 378
      Take the difference   8352
   c. Repeat the process with the digits 8, 3, 5, and 2.
      Largest possible number   8532
      Smallest possible number   − 2358
      Take the difference   6174

After doing taking the difference three times, we get the magic digits 1, 4, 6, and 7.

2. Repeat exercise 1 with the digits 1, 5, 6, and 8. Notice that the magic digits 1, 4, 6, and 7 appear after taking the difference seven times.
3. Study exercises 1 and 2. Start with four digits. Find the four digits with which you must start to get the magic digits 1, 4, 5, and 6 after taking the difference one, two, four, five, and six times.

Activity 9

Use 5, 3, +, and − to construct the numbers 1 through 20.

1 = _____	11 = _____
2 = _____	12 = _____
3 = _____	13 = _____
4 = _____	14 = _____
5 = _____	15 = _____
6 = _____	16 = _____
7 = _____	17 = _____
8 = _____	18 = _____
9 = _____	19 = _____
10 = _____	20 = _____

Activity 10

Press each key once to obtain the answer indicated by the calculator.

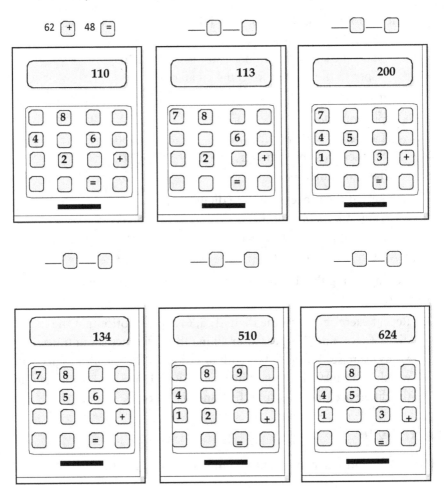

Press each key once to obtain the answer indicated by the calculator.

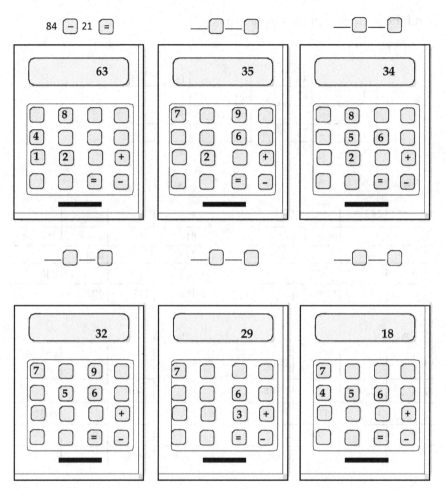

Activity 11: Addition Puzzles

Find the number that goes in each box.

Puzzle 1

+	36		15
11	47	40	
24		53	39
	78		
18	54		33

Puzzle 2

+	243		712
165		308	
	337		806
415			
		659	

Puzzle 3

+				
81	146			354
216		628		
573				846
192	257	604		

Puzzle 4

+			2816
714	1020		3530
		3434	4163
312		2399	
			3335

Puzzle 5

+		614	
	1403		
156			8632
		900	
3948	4748		

Activity 12

Use a calculator to fill in the boxes.

```
     365            1,476          91,263
    +127             +863         +27,619
   [      ]         [      ]      [        ]
```

```
     726            6,432          17,267
  +[     ]        +[     ]       +[      ]
     917            6, 783         88.888
```

```
   [      ]        [      ]       [        ]
    +706           +1,276         +70,261
     999            6,783          87,654
```

```
      73             174           5,091
      26              89          [       ]
    +192         +[      ]          +725
   [      ]           563          6,783
```

Activity 13

1. Round off these numbers to the nearest hundred.
2. Add the rounded-off numbers using a calculator.
3. Check with the answer given below.
4. Use the number line if you need help.

a.

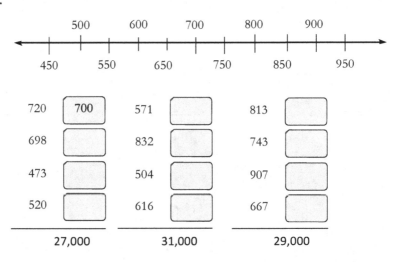

720	**700**	571		813	
698		832		743	
473		504		907	
520		616		667	
	27,000		31,000		29,000

b. Round off these numbers to the nearest thousand and check your answer with a calculator.

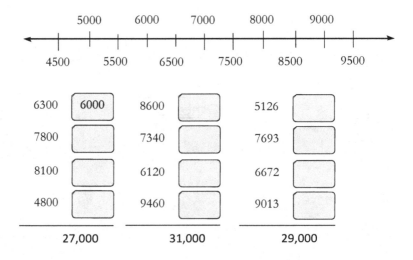

6300	**6000**	8600		5126	
7800		7340		7693	
8100		6120		6672	
4800		9460		9013	
	27,000		31,000		29,000

Estimation

Estimation, besides being an everyday activity, offers an opportunity to identify whether students are learning the concepts. In this section, we include examples of addition and subtraction estimation problems.

Mental math

Approximate the answer and round it off to the nearest ten.

1. 81 +47	**2.** 69 + 23	**3.** 54 −17	**4.** 152 −78	**5.** 119 − 32

6. 29 + 54 ⟶ _____   **7.** 137 − 66 ⟶ _____

8. 72 − 87 ⟶ _____

Approximate the answer and round it off to the nearest hundred.

9. 382 +159	**10.** 765 +138	**11.** 465 −211	**12.** 588 −165	**13.** 1429 −574

14. 1215 − 617 ⟶ _____   **15.** 1502 − 935 ⟶ _____

16. 279 − 405 ⟶ _____

Approximate the answer and round it off to the nearest ten.

17. 17392 +8612	**18.** 7493 +2610	**19.** 5044 −3897	**20.** 12266 −6774	**21.** 11806 −2924

22. 14293 − 5898 ⟶ _____   **23.** 8544 + 7990 ⟶ _____

Approximate the answer and round it off to the nearest hundred.

24. 17392 +8612	**25.** 7493 +2610	**26.** 5044 −3897	**27.** 12266 −6774	**28.** 11860 −2924

29. 14293 − 5898 ⟶ _____   **30.** 8544 + 7990 ⟶ _____

Use mental math to find the cost of buying certain items.

STORE

Pencils	10¢	Protractors	79¢
Notebooks	50¢	Pens	75¢
Erasers	20¢	Scissors	98¢
Rulers	49¢	Glue	69¢

Example:

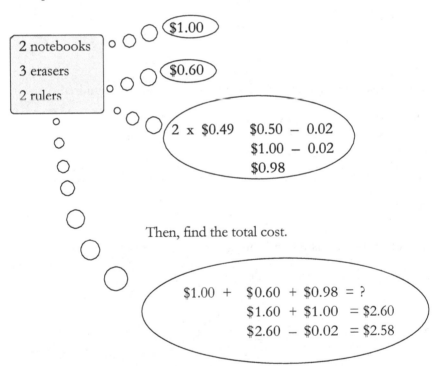

2 notebooks
3 erasers
2 rulers

$1.00

$0.60

2 x $0.49 $0.50 − 0.02
            $1.00 − 0.02
            $0.98

Then, find the total cost.

$1.00 + $0.60 + $0.98 = ?
$1.60 + $1.00 = $2.60
$2.60 − $0.02 = $2.58

Using mental math verify each total. Circle **Too much, Not enough,** or **Correct.**

1. 2 notebooks
   3 pencils
   1 pair of scissors
   You pay $2.28

   Too much
   Not enough
   Correct

2. 2 pencils
   2 pens
   2 rulers
   You pay $3.00

   Too much
   Not enough
   Correct

3. 1 glue
   1 ruler
   1 pair of scissors
   You pay $1.16

   Too much
   Not enough
   Correct

4. 6 pencils
   4 notebooks
   2 erasers
   You pay $2.90

   Too much
   Not enough
   Correct
   Correct

5. 20 pencils
   10 notebooks
   10 erasers
   You pay $10

   Too much
   Not enough
   Correct
   Correct

6. 2 rulers
   1 pair of scissors
   1 pen
   1 notebook

   Too much
   Not enough
   Correct
   Correct

Apples: 23¢
Cookies: 67¢ per box
Eggs: 77¢ per dozen
Milk: 43¢ per liter
Bread: 53¢ per loaf
Onions: 33¢ per kg
Cheese: 87¢ per package

Instructions: Write yes or no.

Example: Jorge went to Happy Valley Market to buy a box of cookies and two apples. He had $1.00. Was that enough? *No.*

Justification: One apple is approximately 20¢ and two is 40¢. Cookies are approximately 70¢. Hence, 40 + 70 = 110 or $1.10. So, $1 is not enough.

1. Sally went to Happy Valley Market to buy a dozen eggs and two loaves of bread. She has $2.00. Will that be enough money?
2. Jerry went to Happy Valley Market with $3.00. He wants to buy a package of cheese and two boxes of cookies. Is that enough money?
3. Joan bought a bag of potatoes for $1.79 and a loaf of bread. She had $2.00. Was that enough money?
4. Henry bought a box of cereal for 99¢ and a package of cheese. He had $2.00. Was that enough money?
5. Frank has $2.25. He wants to buy two liters of milk, a package of cheese, and one kilogram of onions. Does he have enough money?
6. Janet has $2.75. She wants to buy a bottle of juice for 89¢ plus two dozen eggs and a loaf of bread. Does she have enough money?

Patterns

Activity 1 — group activity: What is my rule?

A teacher thinks of a rule to generate triplets of numbers and shows three numbers produced by the rule. The students' objective is to guess the rule. If they guess an incorrect rule, they get −1 point.

To guess the rule, students present a group of three numbers to the teacher, who says if they are generated by the rule.

1, 4, 7

The following triplets satisfy the rule:

2, 5, 8
3, 6, 9
40, 43, 46

The following are examples of triplets that do not satisfy the rule but that can come from a rule that satisfies the triplet 1, 4, 7. Next to each triplet, there is a rule that might generate it.

3, 4, 5   Two odd numbers with an even number in between.
1, 2, 3   Three numbers in ascending order.

Once the teacher has identified students for whom it is easier to generate examples, counter-examples, and rules, the teacher can divide the group into two, where the faster group presents the most difficult rules.

Examples of possible rules

1. Numbers counting by two, by three, by four, by five, and so on.
2. Any three even numbers.
3. Any three odd numbers.
4. Numbers in ascending order.
5. Two odd numbers and an even number between them.
6. Two odd numbers and an even number (in any order).

Activity 2 — game for groups of students

This game is just like the one for individual students, but now there will be several teams. Students will take turns leading their teams to present their answers.

Pascal's pumpkins

1. What numbers go on the last row?
2. Add the numbers in each row. Do you see a pattern? Which one?
3. What other patterns can you find?

CHAPTER 5

MULTIPLICATION AND DIVISION

5.1 Constructing the Meaning of the Operations

As with the learning of addition and subtraction, the development of multiplication and division must be gradual, allowing for student exploration and reflection. The symbols for multiplication and division, as in addition and subtraction, should not be encountered the first time students see situations that model these operations. In fact, the symbolism should arise after students have analyzed situations that exemplify these operations. The steps are similar to the development of addition and subtraction. We begin by allowing a qualitative analysis of situations, beginning with the easiest ones for students (Anghileri, 1989; Lampert, 1986).

5.1.1 *Situations that model multiplication*

Using a topic that is being discussed in class, we introduce a problem that requires mathematical analysis and that can be solved through multiplication. We might be planning a field trip and need to buy two juice bottles and three bags of healthy chips per student. How many juice bottles do we need to buy? How many bags of chips? As we introduce these situations, we also present counting exercises like the following.

How many beads are there? How many cookies? How many lollipops? How many jars?

These exercises (as well as others in diverse contexts involving symmetry, area, combinations, and even fractions!) help to develop the idea of multiplication as repeated addition. Initially, students will individually add the items. In the process, they will develop the ability to count by twos, threes, and so on. This way of counting fosters the learning of the multiplication tables. Then we introduce the representations of repeated additions, this time using the word *times*.

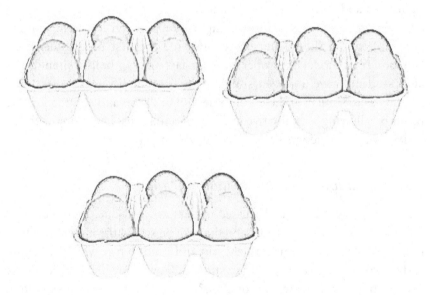

As students learn the meaning of multiplication, it is necessary to start a process of automating the basic combinations.

5.2 Basic Combinations

Once students have some experience in second grade (seven to eight years) with problems similar to those shown above, we start teaching the multiplication tables from easiest to hardest; it is not necessary to learn all of them at once. For example, in second grade, students should learn the tables for two, ten, and five.

The process of learning the basic combinations can be simplified if we help children discover several patterns, like the relationship between the times tables for four and eight. For instance, consider six horses. Each horse has four legs; hence, students can count the legs of six horses.

Likewise, if we now have four mares with their foals, they can now see that the number of legs has doubled.

Thus, if $4 \times 6 = 24$, then 8×6 is twice the previous result — that is, 48.

Another property that supports learning the multiplication tables is the commutative property. Thus, if I know that $4 \times 6 = 24$, I also know that $6 \times 4 = 24$. Knowledge of this property cuts in half the amount of data that students need to learn; hence, they should be made aware of this economy of thought.

5.2.1 *Learning the multiplication tables by association*

When learning the times tables, we can start from the known ones to create our own combinations. For example, the table for five is relatively easy to learn and it can help us learn the tables for four and six. Therefore, multiplying by four — equivalent to adding a number repeatedly four times — is equal to adding the same number one fewer times than when multiplying by five. In general:

$$4 \times n = (\mathbf{5 \times n}) - n.$$

For example:

$$4 \times 6 = (\mathbf{5 \times 6}) - 6 = 30 - 6 = 24,$$
$$4 \times 7 = (\mathbf{5 \times 7}) - 7 = 35 - 7 = 28.$$

In the table for six there is a similar situation, but instead of subtracting we need to add.

$$6 \times n = (\mathbf{5 \times n}) + n.$$

For example:

$$6 \times 7 = (\mathbf{5 \times 7}) + 7 = 35 + 7 = 42,$$
$$6 \times 8 = (\mathbf{5 \times 8}) + 8 = 40 + 8 = 48.$$

Just as from the table for five we can deduce the tables for four and six, we can also deduce the table for three from the table for two.

Another table that is relatively easy to learn is the one for nine, since it can be deduced from the table for ten.

$$1 \times 9 = \mathbf{10} - 1 = 9,$$
$$2 \times 9 = \mathbf{20} - 2 = 18,$$
$$3 \times 9 = \mathbf{30} - 3 = 27,$$
$$\vdots$$
$$9 \times 9 = \mathbf{90} - 9 = 81.$$

Once we know the table for nine, we can deduce the table for eight. The only thing left to do is to memorize the table for seven. Emphasizing this point should be a psychological relief for students (and parents) who think they need to memorize many seemingly unrelated facts while learning their multiplication tables.

5.2.2 *Different rates of learning the tables*

Students who have not learned all the tables are not necessarily limited in the development of other areas of numerical calculations. We can still teach some algorithms and practice with exercises that focus on the tables they know. For instance, suppose students only know the tables for one, two, and five. Then we can teach the multiplication algorithm by permuting the digits of a multiplicand formed with one, two, or five.

387	459	2,853
x 5	x51	x125

5.3 Diverse Problems Represented with the Same Operation

As in addition and subtraction, there are different situations that illustrate multiplication, some of which are naturally easier than others. Multiplication and division introduce new elements that can be challenging for students (Graeber and Campbell, 1993; Greer, 1992; Steffe, 1994). For example, unlike addition and subtraction, multiplication and division transform the referent. In other words, the elements in a sum or difference, including the result, represent the same thing. For example,

$$5 \text{ oranges} + 8 \text{ oranges} = 13 \text{ oranges},$$
$$5 \text{ oranges} + 8 \text{ apples} = 13 \text{ fruits}.$$

However, in multiplication and division, each element (multiplier, multiplicand, and product, or dividend, divider, and quotient) can represent different things in the same operation. For instance,

$$5 \text{ oranges per dollar} \times 3 \text{ dollars} = 15 \text{ oranges},$$
$$10 \text{ inches} \times 6 \text{ inches} = 60 \text{ square inches},$$
$$12 \text{ apples} \div 4 \text{ children} = 3 \text{ apples per child}.$$

Thus, it is necessary to offer students opportunities to explore and understand the situations signified by multiplication and division. We should introduce situations at different levels and allow students to solve them using their strategy of choice. Seeing the invariants in these situations is part of student development.

Here, are several multiplication situations, in order of difficulty:

(1) A corner shop sells 5 lollipops in a bag. Maria bought 3 bags. How many lollipops did she buy?

(2) Sonia has 5 small chocolate bars. Carmen has 3 times as many chocolate bars as Sonia. How many chocolate bars does Carmen have?

(3) At a birthday party, each child received 3 bags of candy. There were 5 candies in each bag. How many candies did each child get?

(4) Rick has 5 shirts and 3 pairs of pants. In how many different ways can Rick get dressed, if he must wear a shirt and pants?

Research Questions

(1) How do students originally understand the different situations represented by multiplication? What kind of difficulties do they face?

(2) What types of activities allow teachers to understand how students interpret these (or similar) problems? Drawings? Comparison?

(3) What types of activities and reflections help students understand these different representations? Drawings? Comparison? Analogies?

The strategies used by children to solve these problems help us to see where they are struggling and to appreciate their different representations of multiplication. In fact, student representations show us a path for concept development, since we can use the representations of those with a better understanding to support the learning of those who do not understand. An example of this situation is presented by Quintero (1988), who offers several representations made by children when solving problems similar to number 4. The representations of one child may serve as intermediate representations to achieve the understanding of the concept.

5.4 Multiplication Algorithm

Similarly to addition and subtraction, the elements for students to construct the multiplication algorithm are developed through working on multiplication problems.

(1) **Basic combinations**

The first multiplication problems can be about some basic combinations, starting in second grade (seven to eight years).

Can you write these multiplications in another way?

_____ *times* 2 is _____

_____ **x** 2 = _____

In second grade, we start with combinations of two, five, and ten. In third grade (eight to nine years), we strengthen those combinations and introduce combinations of three and six, and four and eight. In fourth grade (nine to ten years), we complete this process by introducing the last basic combinations, those of seven.

(2) **Multiplying multiples of ten and one hundred by one-digit numbers**

In third grade, we introduce problems where students have to multiply a one-digit number by multiples of ten. For example:

$$5 \times 40 \qquad 9 \times 40 \qquad 3 \times 60.$$

These problems show the relationship between multiplying by ones and tens.

$$5 \times 4 = 20 \qquad 5 \times 40 = 200,$$
$$6 \times 3 = 18 \qquad 6 \times 30 = 180.$$

(3) **Multiplication of two-digit numbers by one-digit numbers**

In the second semester of third grade, we introduce problems that require multiplication of two-digit numbers by one-digit numbers. These problems promote that students multiply ones by tens. For example:

How many paint cans are there?

7 x 12

7×12

(4) **Multiplication of two-digit numbers by tens**

In fourth grade (nine to ten years), we introduce multiplication of two-digit numbers by tens.

(5) **Multiplication of two-digit numbers by two-digit numbers**

After students have experimented with multiplication of two-digit numbers by tens, we introduce multiplication problems of two-digit numbers by two-digit numbers. To explain the algorithm, students have had the opportunity to work with some of its elements.

5.5 Division

Just as with addition and subtraction, there is a parallel development for multiplication and division. Once students know the meaning of multiplication, we start working with the meaning of division. Then, working with basic combinations, the algorithms develop in parallel, first multiplication and then division. There is no need to wait until students learn the entire multiplication algorithm to start working on division; instead, we integrate multiplication and division problems.

5.5.1 *Constructing the meaning of the operation*

There are two situations represented by division:

— The action of separating or distributing a total into equal parts.

Mary has 30 cookies. She wants to separate them in 5 bags with the same amount in each bag. How many cookies will be in each bag?

— The action that seeks to determine the number of times that a certain quantity is contained in a larger one.

Mary has 30 cookies. She wants each bag to have 5 cookies. How many bags can she make?

We will call the first type of problem **division by distribution**, and the second **division by ratio.**

Starting in kindergarten, students can solve simple problems like these. González[1] (1996) studied the strategies used by kindergarten, first grade, and second grade students working on these types of problems.

---

[1] González's research shows us how children, starting in the early grades, can develop strategies to solve division problems. When comparing children's strategies in different grades, the study also shows how these strategies become more efficient as students mature. It would be interesting to see similar research projects with students that already understand multiplication. What strategies do students use to solve division problems once they know multiplication but have not been introduced to division?

Starting in second grade(seven to eight years), we introduce distribution problems.

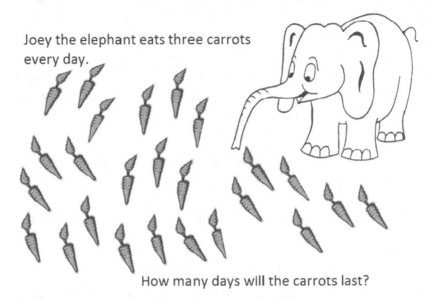

Joey the elephant eats three carrots every day.

How many days will the carrots last?

In this process, we present examples where objects can be distributed evenly as well as cases where this cannot be done. This allows students to encounter the notion of a remainder, informally and intuitively at first. We also integrate problems that present division by ratio.

5.5.2 *From situations to symbols*

Once students have worked with division problems, we can introduce the symbolism for division. As with other operations, symbols are introduced along with illustrations, until we introduce problems with numbers only.

5.5.3 *Studying the properties of the operation between numbers*

In the case of division, as with subtraction, it is not necessary to memorize all the basic combinations, but rather to be able to relate this operation to multiplication. The study of the relation between these two operations starts in third grade (eight to nine years).

$$4 \times 2 = 8 \qquad 8 \div 4 = 2,$$
$$3 \times 5 = 15 \qquad 15 \div 3 = 5.$$

5.5.4 *Several problem structures can be represented with the same operation*

As with multiplication, there are different situations that are modeled with division. We introduce them at different levels and allow students to solve them using their preferred strategies. A part of child development requires seeing the invariant in these situations, for example, that these situations adjust to the model of division.

Below, we present some situations that model division.

Case 1

Division by distribution

Division by ratio

A store sells candies in bags. Jamie has 30 candies and she wants to make 6 bags. How many candies go in each bag?

A store sells 5 candies per bag. Jamie has 30 candies. How many bags can she make?

Case 2

Division by distribution

Division by ratio

Laura has $18. This is three times the amount that Rose has. How much money does Rose have?

Laura has $18. Rose has $6. How many times greater is Laura's amount compared to Rose's?

Case 3

Division by distribution

Division by ratio

A store gave away 3 bags of candy canes to every child. Each child got 18 candy canes. How many candy canes were there in each bag?

A store gave away 18 candy canes to every child. The canes were in bags of 6 each. How many bags did each child get?

Case 4

Combining his pants and shirts, Juan has 18 different ways to dress. If he has three pairs of pants, how many shirts does he have?

The strategies used by children to solve these problems help us see where they are struggling and appreciate their different representations of division. As with multiplication, student representations show us a path for concept development, since we can use the representations of those with a better understanding to help other children advance in the learning sequence.

Research Questions

(1) How do students originally understand the different situations represented by division? What kind of difficulties do they face?

(2) What types of activities allow teachers to understand how students interpret these (or similar) problems? Drawings? Comparison?

(3) What types of activities and reflections help students understand these different representations? Drawings? Comparison? Analogies?

5.5.5 *Division algorithm*

The development of the division algorithm follows a very similar process to that of the multiplication algorithm.

(1) Basic combinations
The first division problems revolve around basic combinations.

$$12 \div 4 = 3.$$

(2) Division of multiples of ten or one hundred by one-digit numbers
In third grade (eight to nine years), shortly after performing this same task with multiplication, we introduce problems in which students have to divide multiples of ten by one-digit numbers. For example:

$$60 \div 3.$$

If unable to see it by themselves, during these problems students should be guided to see the relationship between $60 \div 3$ and $6 \div 3$.

(3) Division of two-digit numbers by one-digit numbers
In third grade, we introduce problems in which we have to divide two-digit numbers by one-digit numbers. By doing this, we foster division of tens by ones.

(4) Division between two-digit numbers: multiples of ten
In fourth grade (nine to ten years), we introduce division between two-digit numbers that are multiples of ten. By doing this, we establish a relationship with the division of the digits corresponding to the tens' place.

$$5 \div 1 \rightarrow 50 \div 10.$$

(5) Division between two-digit numbers
In fifth grade (ten to eleven years), we present problems in which students divide two-digit numbers. We ask students to develop their strategy. We use the discussion of their strategies as a way to introduce the algorithm.

5.5.6 *Development of multiplication and division*

Unlike the traditional curriculum, in which multiplication is presented first and then division, we propose that these two operations be taught simultaneously. At each stage, we present multiplication first and then division, without having to wait for the development of every multiplication skill before moving on to division.

We can describe the development of these operations in the following way:

Multiplication: meaning of operation $\rightarrow$ situations to symbols $\rightarrow$ algorithm

Division: meaning of operation $\rightarrow$ situations to symbols $\rightarrow$ algorithm

5.6 Mental Arithmetic

As with addition and subtraction, we should encourage students to solve mental arithmetic problems with multiplication and division. For example, you may ask them to mentally solve 28×3. When working on this problem with a group of students, we found the following three ways to think it through:

Joe said:
$$28 \times 3 = 20 \times 3 + 8 \times 3 = 60 + 24 = 84.$$
Linda said:
$$28 \times 3 = 25 \times 3 + 3 \times 3 = 75 + 9 = 84.$$
Miguel said:
$$28 \times 3 = 30 \times 3 - 2 \times 3 = 90 - 6 = 84.$$

All of these are correct ways to solve the problem. After students know the times tables and certain properties of multiplication and division, we should include mental arithmetic problems.

5.7 Estimation

As with addition and subtraction, once students have worked with the algorithms and mental arithmetic in multiplication and division, we introduce estimation problems involving these operations. The process is similar to the one explained for addition and subtraction. Although in this case, students already recognize the function of estimating.

5.8 Calculators

When working with problems that require multiplication or division calculations that take a considerable amount of time, we integrate the calculator. With these operations, as with addition and subtraction, calculators have several functions. Calculators are learning tools that support teachers in their creation of situations that foster reflection on the structure of our number system. Finally, they are useful for the solution of problems and for finding patterns.[2] For example, we ask students to choose any number and multiply it by ten several times.

$$23 \times 10 = 230,$$
$$230 \times 10 = 2300,$$
$$2300 \times 10 = 23000.$$

What are their observations? Each time that we multiply by ten, a zero is added to the number. Why does that happen? This question makes students reflect on the structure of our number system. Thus, we ask: "What do you think will happen when dividing by ten several times?"

$$23 \div 10 = 2.3,$$
$$2.3 \div 10 = 0.23,$$
$$0.23 \div 10 = 0.023,$$
$$0.023 \div 10 = 0.0023.$$

---

[2]To the surprise of the second author, in 1996 he walked into the office of Princeton's mathematician John Conway. He was amused to find him with two graduate students using hand calculators to look for patterns in a problem they were working on the board.

Notice that when we divide by ten, the decimal point moves to the left. If the original number chosen for multiplication by ten had decimal digits, would the decimal point move to the right when it is multiplied by ten?

References

Anghileri, J. 1989. "An investigation of young children's understanding of multiplication", Educational Studies in Mathematics, 20, 367–385.

González Jasso, D.M. 1996. "Estudio cualitativo sobre la formación del concepto de división aritmética en la niñez", Masters thesis, University of Puerto Rico, Río Piedras.

Graeber, A.O. and Campbell, P.F. 1993. "Misconceptions about multiplication and division", Arithmetic Teacher , 40, 408–411.

Greer, B. 1992. "Multiplication and division as models of situations", in Grouws, D.A. (ed.), Handbook of Research on Mathematics Teaching and Learning. New York: Macmillan Publishing Co., pp. 276–295.

Lampert, M. 1986. "Teaching multiplication", Journal of Mathematical Behavior, 5(3), 241–280.

Quintero, A.H. 1988. Representaciones en la enseñanza de las matemáticas. San Juan: Editorial de la Universidad de Puerto Rico.

Steffe, L.P. 1994. "Children's multiplying schemes", in Guershon, H. and Confrey, J. (eds.), The Development of Multiplicative Reasoning in the Learning of Mathematics. Albany, N.Y.: State University of New York Press, pp. 3–39.

CHAPTER 6

FRACTIONS, DECIMALS, AND PERCENTAGES

6.1 Introduction

The idea of a rational number is one of the most complex and important ideas introduced in elementary school (Behr *et al.*, 1983). Research in several countries shows that the concept of fractions or rational numbers presents a high degree of difficulty for students (Gravemeijer, 1994; Karplus *et al.*, 1983; Streefland, 1991). Unlike integers, fractions refer to a **relation**, not a quantity. For example, if we have several objects, say four rectangles:

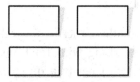

and take one of them, the question "What fraction is ▢?" is meaningless, because to speak of fractions we need to have a relation between two quantities. In this example, the rectangle could be an integer if it is defined as a unit, or it could be a half or a quarter, depending on what is defined as a unit in that context.

This property of fractions is difficult to grasp. It becomes more difficult because fractions can be interpreted in different ways and can be represented by using several numerical formalisms. For example:

(1) As a **comparison** of part to whole (1/2 can represent half of a cake).
(2) As a **decimal** (for every 10 people, 2.5 are prone to develop heart conditions before they reach 30 years of age).
(3) As a **ratio** (2 red balls, in a collection of 5 balls, represent the same proportion as 4 red balls in a collection of 10).
(4) As a **division** (we can interpret the fraction 21/7 as the division of 21 by 7 — that is, 3).
(5) More generally, as an **arithmetic operation** (e.g. $(2 - 3)/(4 + 5)$).
(6) As a **measurement** of discrete or continuous quantities (the distance between the school and the house is 2/3 of the distance between the house and the town square).

As if this plethora of interpretations was not enough — increasing the complexity of learning this concept — fractions allow for at least three numerical formalisms for their representation:

(1) Fraction formalism or integer ratio (these include proper and improper fractions, as well as the so-called mixed numbers: 1/2, −4/5, 3/7, 8/3, 21/4, etc.)
(2) Decimal formalism (0.25, 0.33, −0.5, 0.2224, ... , etc.)
(3) Division formalism ($10 \div 2$).

As we have stated, in mathematics in general, many of the challenges faced by students when working with fractions are due to insufficient experience to give meaning to newly acquired knowledge. As with the concept of number and operations with numbers, the concept of fraction requires a gradual developmental process (Mack, 1990). Our first task is to offer contexts that provide fractions with meaning — real situations that allow for the development of analogies to help students interpret fraction operations and relations.

6.2 Context

We have emphasized that before introducing numerical symbols, it is important to offer representations that give meaning to those symbols.

Fractions, like natural numbers, arise in different contexts and with different meanings. We should develop these diverse notions and the connections between them (D'Ambrosio, 1994).

Different interpretations of fractions have several degrees of difficulty. Even in just one interpretation there might be problems of different difficulty levels. Thus, we should start with the simplest interpretations and gradually introduce more complex contexts and situations. Moreover, we should not wait until all interpretations of fractions are introduced before we start working with symbols, operations, and relations; rather, we can work concurrently, as shown in the following table.

Fractions in Context		
Grade (Age in Years)	**Context**	**Operations and Relations**
Kindergarten (5–6)	Dividing in equal parts	
First (6–7)	Dividing in equal parts	Comparing
Second (7–8)	Dividing in equal parts	Comparing
Third (8–9)	Dividing in equal parts Measurement Unit ratio	Comparing Introducing symbolism Addition with like denominators
Fourth (9–10)	Deepening previous contexts and introducing ratio and division	Comparing Reinforcing symbolism Addition and subtraction with like denominators Multiplication and division ideas
Fifth (10–11)	Deepening previous contexts	Relating interpretations Addition and subtraction Equivalent fractions
Sixth (11–12)	Deepening previous contexts	Relating interpretations Estimating with multiplication and division

6.2.1 *Dividing in equal parts*

The part–whole idea is the fraction interpretation best understood by students. From a relatively young age, students confront situations in which it is necessary to divide a whole into equal parts (Empson, 1995). For instance, when cutting an apple or a candy bar into equal parts for two or four siblings, parents talk about cutting it in halves or quarters, and so on. Starting in second grade (seven to eight years), we can build on this informal knowledge with problems such as the following:

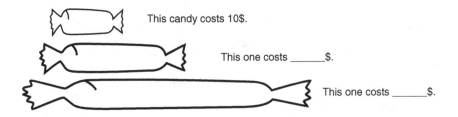

This candy costs 10$.

This one costs _____$.

This one costs _____$.

As this idea matures, we introduce more complex problems to represent fractions as part–whole. Here are some problems that give contexts for the part–whole idea.

How much is the shaded part?

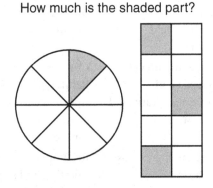

The part–whole idea is a unifying metaphor of the different interpretations of fractions. When working on measurements and starting to calibrate the ruler, we should start by dividing a whole into equal parts. One representation used when introducing the concept of ratio is the part–whole idea.

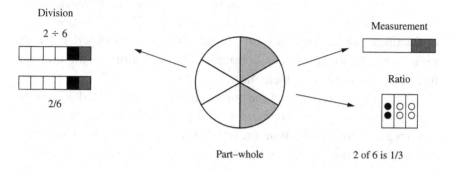

6.2.2 **Measuring**

When measuring, fractions arise naturally.

Activity

Each child is given a paper strip that is as long as the width of a sheet of paper:

We ask them to measure several objects with the strip. Once we start measuring objects with several strips of the same size, there will be cases in which an object cannot be measured exactly in a whole number of strips. You may ask: "What should we do?" Taking the children's suggestions as a starting point, we can guide them until we reach the idea of dividing the strip into fractions: half the strip (1/2), half of half the strip (1/4), etc. We continue this process until we reach a reasonable division for the children's levels, since it is likely to have several reasonable divisions in the same classroom. Some children might understand up to 1/4 while others might get to 1/8.

While leading this activity, we should be aware that understanding that half of a half is a quarter it is not obvious for children. Reaching this understanding requires time. However, we reiterate that it is better to take the time to learn the concepts with meaning than to go faster and learn things without understanding.

The strip is another analogy that supports fraction comprehension at the same time that it allows for the integration of different interpretations of fractions. For instance, as soon as students have more experience with measurements, we can integrate fraction representation on the number line, using the measuring metaphor.

Representing fractions on the number line is not an easy task. Consider the following two errors we found in middle-school and college students when representing fractions on the number line. After describing the error, we analyze the misconception that led to it.

Problem: Graphically represent (on the number line) the fractions 3/4, 5/2, and 5/6

Error 1

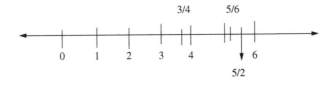

This student does not understand the representation of fractions on the number line. In fact, after making this representation the student said, "I don't understand." When making this representation, he first takes the numerator as the number to locate on the number line. Then, he moves to the right of that number and marks the number line before the next whole number, thinking of the denominator as a unit fraction.

Error 2

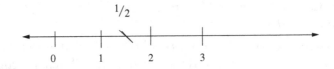

Several (middle-school and college) students made this mistake, which is common in children. They usually understand that one-half lies in the middle of "something," but cannot determine the start of the measuring unit; they start from one instead of zero.

These errors show that students have a clear idea neither about fractions nor of the number line. For this reason, it is important to relate the number line to measurements, and measurements to fractions.

To correct this misconception, we can use the sample exercises to support the comprehension of fractions on the number line. This number line portrays the entire number line as the sum of strips, with each strip equivalent to one unit. Once we understand that each space on the number line represents one unit, we can go on to ask for the representation of, say, 1/2 or 1 2/3. In the first case, it is half of the first unit; the second case is one unit plus 2/3 of the second unit.

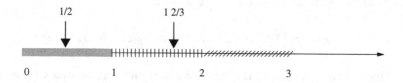

Moreover, representing fractions on the number line reinforces the learning of equivalent fractions.

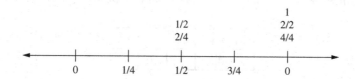

6.2.3 *Ratio*

Ratio is a crucial concept in mathematics, as well as in science and statistics. If we think about the curricula to which we have been exposed, they seldom propose the direct teaching of this concept. Ratio, however, is a difficult concept for students (Lappan *et al.*, 1998). As in many others, it requires a gradual construction.

Research shows that the first type of ratio that students grasp is the ratio–unit concept. This is the ratio that is expressed in fractions such as 4/1, 8/1, 12/1. For example:

> 4 candies in each bag
> 8 oranges per dollar
> 12 children per teacher

From kindergarten to grade 3, we can integrate problems with these types of ratios. Some of the challenges faced by students with this type of ratio are due to language, like understanding that "8 oranges per dollar" equals "8 oranges for each dollar." Once students understand the ratio–unit concept, we can introduce activities with other types of ratios around fourth grade (nine to ten years). For example:

Is it the same to have 2 pizzas per 3 students as 4 pizzas per 6 students?

One way of reinforcing the understanding of more complex ratios is to help students see them in terms of the ratio–unit. One way to guide them in seeing this relationship is by using tables, like in the following problem.

Carmen bought 10 candies for 5¢. How many candies can she get for 8¢?

Money	1¢	5¢	8¢
Candies		10	

If we discover how many candies are sold for 1¢, then it will be easier to know how many are sold for 8¢. If we get 10 candies for 5¢, then we can get 2 candies for 1¢. From this, we can calculate that you can buy 16 candies for 8¢.

Money	1¢	5¢	8¢
Candies	2	10	16

Let us see another example.

Maria went to buy some lollipops. She got 10 for 25¢. How many lollipops can Roberto buy for 10¢?

We make the following table and ask students to try to find the missing values.

Money	5¢	10¢	15¢	20¢	25¢
Lollipops					10

This is a possible strategy:

Step 1

Since we know that 5¢ goes into 25¢ five times, the number of lollipops sold for 5¢ must go into 10 five times — that is, 2.

Money	5¢	10¢	15¢	20¢	25¢
Lollipops	2				10

Step 2

If we get 2 lollipops for 5¢, then for twice that amount we should get twice as many lollipops.

Money	5¢	10¢	15¢	20¢	25¢
Lollipops	2	4			10

In the process of working with these problems, it is important for students to understand that, in ratios, absolute quantities are not as important as the relationship between them. We can introduce activities that work on this idea.

Which lemon juice is more concentrated, one in which we use 2 lemons for every 3 cups of water, or one in which we use 3 lemons for 5 cups of water?

How do we solve this problem? Many students will say that the juice with three lemons has more lemon flavor because it has the juice of more lemons. We point out that while it does have more lemons, it also has more water, which dilutes the flavor. This discussion leads to the idea that in a ratio it is important to analyze the relation between both quantities. Which one has a higher lemon concentration per cup of water?

In this process it is important to understand that the relationship given in a ratio is not additive but multiplicative. Here are some examples of activities that help develop this type of relation.

Activity 1

Which class has better attendance: Mrs. Perez's classroom with 8 of 12 students in attendance, or Ms. Wayne's with 10 of 20 students in attendance?

Activity 2

Problems that deal with different forms of choosing sitting arrangements — or distribution in general — provide another context that leads to the ratio idea (Empson, 1995). These problems also help in analyzing equivalences between fractions. Let us see an example.

There were 8 people invited to an activity and 6 pizzas were ordered. Ana suggested sitting them as follows:

But Luisa suggested sitting them like this:

Which arrangement seems more appropriate? Why? Will the pizzas be distributed equally in both arrangements? What other arrangements are possible in such a way that everyone has equal access to the same amount of pizza?

Activity 3

It is important to relate ratio to the part–whole interpretation. For example, if we have one whole divided into three parts and we take two of them:

Is that equivalent to having the same whole divided into six parts and taking four of them?

One way of working with this comparison is the "double line," in which a whole is divided one way above and another way below:

When comparing them, we observe that two out of three is the same as four out of six. We can also use this method to compare fractions. For example, 2/3 is greater than 3/6.

Research Questions

(1) What other analogies support the development of the ratio concept?

(2) In what grade do students understand the idea of the double line?

(3) How effective is this way of reinforcing the ratio concept?

6.2.4 *Division*

Fractions arise when dividing a number by a larger number. For example,

You have three pizzas that must be divided equally between four people. How will you divide the pizzas and how many slices will each person get?

When presenting this problem, we found that children solved it in several ways.

(1) Each pizza was divided into four slices and each person took a slice from each pizza.

In other words, they each got three slices or 1/4 + 1/4 + 1/4.

Here we can try to elicit from students a more efficient way to represent 1/4 + 1/4 + 1/4. As we have previously stated, it is important to present different strategies to solve problems, as well as let students invent their own ways of representing the situation. From the discussion, the representation of 3/4 should arise. That is, each student ate the equivalent of 3/4 of a whole pizza.

(2) First, two pizzas are divided in half and each person takes half a pizza. Then, the other pizza is cut into four slices and each person gets one slice.

Thus, each person gets 1/2 + 1/4 of the pizzas.

Given these first two representations, we may ask whether 3/4 is equal to 1/2 + 1/4.

(3) "Pac-Man" cut.

Three people get 1 − 1/4 and one gets 1/4 + 1/4 + 1/4.

This problem presents a situation that lends itself to the discussion of other fraction properties. Thus, we may ask,

Are these three representations equal, namely 3/4, 1/2 + 1/4, and 1 − 1/4? Why?

This is a way to initiate a discussion about equivalent fractions and begin developing an intuitive idea about them.

6.2.5 *Summary*

We have presented several interpretations of fractions:

- Part–whole.
- Measurement.
- Ratio.
- Division.

For each interpretation, we present several related contexts. As students gain a deeper understanding of fractions, we integrate more complicated contexts, while also presenting activities and analogies that allow for the association of different interpretations. For example, the measuring strip helps to associate measurement with the part–whole idea, given that the strip becomes a unit (a part) with which to measure a whole.

6.3 Symbolism

As we have stated earlier, mathematical symbolism is an instrument that facilitates math tasks. Yet, symbols are not natural for students. As with concepts, we must allow students to adopt standard symbolism after starting from their own representations. The development is very similar to the one previously described for natural numbers and the operations between them.

6.3.1 *From situation to visual models*

Going from a situation that includes the idea of fraction — or the operations between fractions — to a visual model is in itself a developmental step. When building on the visual model, students must leave aside part of the richness of reality and focus on the concept at hand. For example, when solving the following problem:

> *A child has a chocolate bar and wants to divide it in equal parts between five friends. How can he do it?*

We asked a group of children to draw a representation of this situation. Some children drew every detail without paying attention to the factor that required more attention, like this picture:

Others drew a chocolate bar and a structure to divide it:

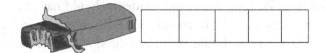

We can introduce the fraction symbolism starting from these drawings.

6.3.2 *From visual models to symbolism*

In some cases, mathematical symbolism has certain relationships with the symbolized objects. Thus, 1/4 has some relation to the fraction (we divide into four equal parts and take one of the parts), or the two parallel lines that represent the equals sign "=," thinking of parallel lines as "equal." In these cases, we should explain the history of the symbolism. In other situations, as with the 0 to 9 numerals, there is no relation — at least not an obvious one — between the symbol and what it represents. In these cases, we must just memorize the symbols and associate them with situations that make to the students.

In terms of fractions, we should start with symbolism for unit fractions (e.g. 1/4, 1/5, 1/6). Once students have some experience with these types of fractions — around fourth grade (nine to ten years) — we introduce situations that require other types of fractions. In all the contexts we have discussed, at one point or another, non-unit fractions arise. For example, in the situation about dividing pizzas, the fraction 3/4 arises. It does not arise directly, but as the addition of 1/4 +1/4 +1/4. This helps students reach a better understanding of the symbol 3/4.

6.4 Improper Fractions

When discussing different interpretations of fractions, we should create situations that lead to mixed numbers and improper fractions. For example, objects that measure 2 1/2 units. In division, we could present a case that calls for dividing five chocolate bars between two children; each gets 5/2 or 2 1/2 bars.

In these contexts we should introduce problems that require changing improper fractions to a mixed number and vice versa. Again, before presenting the rules for these exchanges, students should have the opportunity to work on the changes in context and with meaning. In fact, to the extent possible, the rules should arise from the students themselves.

6.5 Comparing and Ordering Fractions

Once students have a basic idea of fractions, we can start to compare them; this is an iterative task. When we introduce unit fractions, we compare unit fractions. Later, when integrating other types of fractions, we move on to compare them. In our teaching experience, we have found a high level of difficulty in this task. This difficulty, as in other cases, occurs when rules for deciding what fraction is larger than another are presented too quickly, before students get an intuitive understanding of how to reach such conclusions. This leads to students becoming confused in a jungle of rules and, for instance, believing that 2/5 is larger than 3/5 because "the larger the numerator, the smaller the parts." They are using a rule that only applies to the denominator and applying it to the numerator. When those students are asked to represent 2/5 and 3/5 on a number line or other type of drawing, they realize their mistake. Thus, it is important for students to recognize where the rules come from before we enunciate a series of rules.

The following are activities that encourage students to compare and order different representations of fractions:

(1) If divided equally, who gets more sandwich, a group of 3 children who share 2 sandwiches or a group of 4 children who share 3 sandwiches?

(2) Which book is thicker, one that is 3/4 of an inch thick or another that is 1/2 an inch thick?

Comparing fractions leads to equivalent fractions. In some of the previous exercises in this chapter, we saw how "families" of equivalent fractions are created. Moreover, when comparing fractions, we encourage students to find ways to express fractions with the same denominator.

6.6 Equivalent Fractions (From Fourth Grade/Nine Years)

When working on comparing fractions or division, like in the problem about the three pizzas, several ways of labeling the same quantity arise (1/4 + 1/4 + 1/4 and 1/2 + 1/4). In other words, depending on the way that we divide the whole, a given quantity can be expressed using different fractions. Likewise, when comparing fractions, situations arise in which a given fraction is represented in different ways. It is necessary to explore this situation and provide activities that let students discover that fractions have an infinity of representations.

For example, if we divide strips of the same size into halves, thirds, quarters, and so on, and place them one over another, we see that 1/2, 2/4, and 3/6 represent the same quantity. Thus, we call them **equivalent fractions.**

Equivalent fractions will arise in different contexts that will help reinforce this concept. Once students have been exposed to them, we can introduce exercises asking them to find equivalent fractions. Working with these types of exercises will lead students to see patterns from which to deduce a rule for obtaining equivalent fractions.

6.7 Relating the Different Interpretations

As we discuss different interpretations of fractions, we should also relate them. The part–whole idea provides a unifying metaphor. For instance, when working with ratios, we introduce the double line that associates ratios to the part–whole idea. Also, if we have 15 children out of 45, we take 45 as the total and observe that 15 is one-third (a part) of the total.

For measurements, the measuring unit becomes the whole. Thus, when an object measures half a unit, it measures 1/2 of the whole. An object that measures 3 1/4 is calculated as three units and one-quarter of a unit.

When dividing, we can start by fractioning a unit — a chocolate bar, a pizza, an acre of land. When dividing this unit into equal parts between five children, or between four people, each one will get fifths or quarters, respectively. In this case, it is easy to see the relationship between division and the part–whole idea.

When dividing more than one unit, there are several ways to interpret division that arise, as we saw earlier with the pizza example. The interpretation that divides each unit (pizza) and then gives a piece of each unit (a slice of pizza) to each person is best to associate the idea of division with the part–whole idea. In this case, we can clearly see why dividing three by four is equal to 3/4.

6.8 Operations with Fractions

We have emphasized the importance of offering representations of concepts before introducing their symbolism or algorithms. In the case of operations with fractions, as with other operations, students need to develop the following skills:

- Recognizing which operation is better suited to solve a problem.
- Expressing the operation in symbolic form.

Likewise, we have insisted on the need to develop concepts gradually. For operations with fractions, we suggest the following stages.

Grade (Age in Years)	Addition and Subtraction	Multiplication	Division
Third (8–9)	Like denominators	—	—
Fourth (9–10)	Whole numbers and fractions	Interpretation	—
Fifth (10–11)	Different denominators	Interpretation	Interpretation
Sixth (11–12)	Deepen concepts	Algorithms	Algorithms

For each operation we should:

- Start by introducing problems in which students interpret and see the need to use the given operation.
- Develop the algorithm from this situation.
- Offer new situations to practice the algorithm.

6.8.1 *Situations representing the operation*

(1) Addition and subtraction of unit fractions with the same denominator

When studying the different contexts that lead to fractions, we can introduce situations that require the addition or subtraction of unit fractions with the same denominator. This process develops the idea of these operations while also establishing the relationship between the different fractions and the unit fraction. For example:

During recess, five friends decided to share their snacks. Three of them had a chocolate bar. If they divided them equally, how much chocolate did each of them get?

As with the pizza case, this problem can be solved in different ways. We will show the one that more easily leads to the discussion of addition of unit fractions.

The children who had chocolate bars divided their bars into five pieces and each child takes a piece from each bar. Thus, each child gets 1/5 + 1/5 + 1/5. How much is this?

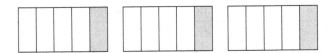

When solving this problem, students learn to add fractions while also learning that 3/5 is the same as 1/5 + 1/5 + 1/5.

(2) **Addition and subtraction of fractions with like denominators**

Once students have some experience with unit fractions, we introduce addition and subtraction situations that are solved with fractions of like denominators. For example:

> *During snack time, Luis gave 2/5 parts of his chocolate bar to Ricardo. Later, Ana gave Ricardo 2/5 of her bar. How much chocolate did Ricardo get? How much was left for Luis? How about Ana?*

(3) **Addition and subtraction of fractions and whole numbers**

Once students understand addition and subtraction of fractions with like denominators, we introduce problems with fractions and whole numbers. For example:

> *Luis has a chocolate bar. Mary gives him one-quarter of hers. How much does he have now?*

(4) **Addition and subtraction of fractions with different denominators**

We introduce problems and let students use their own strategies to solve them. Student strategies are used to help construct the algorithm.

> *Enrique, Antonio, and Victor met during recess. Enrique and Antonio had*
> *a chocolate bar each. Both decided to give a piece of their bar to Victor, who*
> *did not have any. Enrique gave him 1/2 a bar and Antonio gave him 1/4.*
> *How much chocolate did Victor get? Who ended up with the most chocolate?*
> *Who ended up with the least?*

We begin with problems in which one of the denominators is a multiple of the other; then, we introduce denominators that are not multiples of each other. We let students solve the problem by their own methods. Once students have worked on several problems, we discuss their different strategies. If the following strategy does not arise, we can present it as a strategy used in another classroom.

(a) Represent quantities.

(b) Look for a fraction with a common denominator in which both fractions can be represented.

Expressing all fractions with the same denominator allows us to add, subtract, and compare them.

After several problems, we ask students if they can find a way to add and subtract fractions without having to draw picture. The algorithm for addition and subtraction should arise from this exchange. Once this happens, we present exercises to help practice the algorithm.

6.8.2 *Interpretation of multiplication and division*

Usually, rules for multiplication and division of fractions are introduced without explaining their meaning. Our experience informs us that even some teachers do not understand the meaning of these operations. Too often, teaching starts directly with the algorithm. In fact, many adults are surprised that multiplying fractions between 0 and 1 yields smaller results than the two factors.

When introducing multiplication and division, we should present situations that exemplify these operations, as we do with addition and subtraction. While working on addition and subtraction problems with different denominators, we present situations that require the interpretation of multiplication and division of fractions. For example:

> *Bernie bought half a box of pencils. If the box had 20 pencils, how many did he buy?*
>
>> *Mercedes is baking a cake. She decided to make only half of the recipe. If the recipe asks for 1/4 cup of sugar, how much sugar will she need?*

Challenge!

> *George gave John 1/2 of his candies.*
> *John gave Ramon 1/2 of the candies that George gave him.*
> *Ramon gave Margarita 1/2 the candies that John gave him.*
> *If Margarita got 6 candies, how many candies did George have at first?*

6.8.3 *Starting from a situation to develop the algorithm*

After working on problems that can be represented by multiplication, we point out that they are modeled by multiplication, explaining, for instance, that $1/2 \times 1/4$ means half of one-quarter.

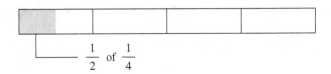

Observe that, in multiplication, the role of each fraction is different; we can take either one as the quantifier and the other one as the amount to be quantified. For this example, let us take 1/2 as the quantifier.

This fraction plays the role of a
quantifier— that is, it tells us what part of
the other quantity we want. In this case,
half of a fourth.

1/2 x 1/4

This fraction plays the role of
the amount to be quantified.

As with addition and subtraction, we present multiplication problems by stages:

- Whole numbers for fractions:
  1/2 of a box with 20 pencils.

- Fractions by fractions:
  1/2 of 1/4 of a cup,
  2/3 of 3/4.

Division

In division, 1/2 ÷ 1/4 means: How many quarters go into one-half? Once students have practiced these problems, we begin looking for regularities that lead to the development of algorithms.

Multiplication algorithm

(1) To determine what 1/2 × 1/4 is, we have to find half of a quarter. That is, we have to divide one-quarter into two equal parts and choose one of these parts. This is equal to saying that the number of parts in

which the whole was divided, four in this case, is multiplied by two. Thus, it will be 4 × 2 = 8 parts. This means the denominator of the product will be 8.

(2) Originally we had one-quarter that was divided by two to be converted into two-eighths. Because we want half of a quarter, we choose one of the two-eighths we obtained by dividing the quarter. That is to say, the *numerator* of the product is

$$1 \times 1 = 1.$$

(3) Thus, the rule to multiply fractions is this: The product of two fractions is a fraction whose *numerator* is the product of their numerators and whose *denominator* is the product of their denominators. In our example, we have:

$$\frac{1}{2} \times \frac{1}{4} = \frac{1 \times 1}{2 \times 4} = \frac{1}{8}.$$

Let us see two graphical representation of this example.

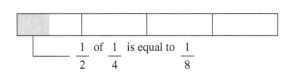

$\frac{1}{2}$ of $\frac{1}{4}$ is equal to $\frac{1}{8}$

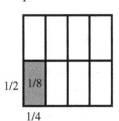

Let us see a second example of multiplication of fractions. Suppose we want to do

$$3/5 \times 1/2.$$

Like in the second type of diagram above, we can divide a square horizontally into fifths and vertically into halves.

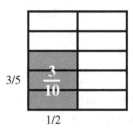

To recap, when multiplying fractions, we multiply the numerators together and do likewise with the denominators. This method clearly illustrates the idea of "multiplication as repeated addition" in the context of fractions; often neglected or misunderstood by researchers and teachers.

Exercises and problems

Once students understand the rules of multiplication and division, we proceed to offer exercises and application problems. While students should know how to multiply and divide fractions, we should not excessively complicate this assignment. We recommend simple exercises that can be easily approached with the square method presented above to do $3/5 \times 1/2$. This will give students a deep understanding of the structure of multiplying fractions. Besides, the most important thing is for students to understand the interpretation of these exercises, given that calculators can execute algorithms. To verify that the calculator results are reasonable, however, it is important to be able to estimate results.

6.9 Decimals

Decimal numbers are the way of expressing fractions in our decimal system. Historically, fractions arose before the development of the decimal system. Thus, there was one way to write integers and another to write fractions. Once the decimal system was developed, both forms of expressing rational numbers were unified into one system. When working with fractions and decimals, we should develop the relationship between those two concepts. Initially, however, we discuss each of them separately.

6.9.1 *Contexts for decimals*

Money

The teaching of decimals should start by using our monetary system, one with which children are in contact from a very early age. Around third grade (eight to nine years), we introduce the representation of money using decimals (i.e. $12.35). In this case, we explain that the decimal place is occupied by cents. The first problems with these numbers will be related to money and shopping.

Measurement

Measurements also lead to decimals. Originally, when calibrating a ruler, we use fractions: 1/2, 1/4, 1/8, and so on. Then we divide the unit into ten and, from there, we can introduce the metric system. This process allows us to work on the representation of decimals on the number line by associating them with fractions.

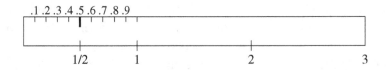

This representation lets us see that 0.5 = 1/2.

Once students have worked extensively with decimals — associating them with measurements — we introduce exercises that represent decimals on the number line. For example, represent 0.4, 0.25, 0.3, or 0.01 on the number line.

| 0 | 0.1 | 0.2 | 0.3 | 0.4 | 0.5 | 0.6 | 0.7 | 0.8 | 0.9 | 1 |

6.9.2 *Operations with decimals*

While working with different contexts, we compare decimals and their operations (Resnick *et al.*, 1989). Rules for addition, subtraction, multiplication, and division should arise from the students' experience in context. In fourth grade (nine to ten years), we begin with the addition and subtraction of decimal numbers, associating them with money. We explain that we add cents with cents, dimes with dimes, dollars with dollars, etc. We note that if we line up the decimal points, everything falls into place:

$$
\begin{array}{r|r}
23 & 12 \\
+\,56 & 34 \\
\end{array}
$$

The subtraction process is similar to the addition process.

In fifth grade (10–11 years) we start comparing decimals, still associating them with money, as well as multiplying decimals by integers. Somewhat later, in the same grade, we introduce division of decimals by integers.

In sixth grade (11–12 years), we expand on multiplication and division to decimals by decimals. To explain the multiplication rule, we interpret decimals as fractions. We write decimals in fractional form, then we multiply fractions and, finally, write the fractional result in decimal form. Here are some examples:

(1) $0.2 \times 0.1 = \frac{2}{10} \times \frac{1}{10} = \frac{2 \times 1}{10 \times 10} = \frac{2}{100} = 0.02$

(2) $0.8 \times 0.15 = \frac{8}{10} \times \frac{15}{100} = \frac{8 \times 15}{10 \times 100} = \frac{120}{1000} = 0.120$

(3) $0.32 \times 4 = \frac{32}{100} \times \frac{4}{1} = \frac{32 \times 4}{100 \times 1} = \frac{128}{100} = 1.28$

(4) $0.25 \times 1.5 = \frac{25}{100} \times \frac{15}{10} = \frac{25 \times 15}{100 \times 10} = \frac{375}{1000} = 0.375$

Multiplication of the numerators corresponds to multiplying the two decimal numbers — as if they were integers — disregarding the decimal point.

To determine the number of decimal places of the product, we observe the product of the denominators. When multiplying the denominators, we are multiplying two powers of ten. The product of two powers of ten is another power of ten with an equal number of zeros to the sum of the zeros of both factors. This is the same as saying that *the number of decimal places of the product is equal to the sum of the decimal places of the factors*. This leads to the rule of multiplication of decimals:

(1) Multiply both decimal numbers as if they were integers and place a decimal point at the end.
(2) Add the decimal places of the two multiplicands.
(3) In the number obtained in step 1, move the decimal point — from right to left — the number of places given by step 2. The resulting number will be the product of the two original numbers.

Students should do operations with decimals just to learn the rules, not to become adept at multiplying excessively long numbers; those should be worked with calculators.

6.9.3 *Relating decimals to other fraction representations*

During fifth and sixth grade (10–12 years), we introduce problems that require associating decimals with fractions and percentages. For example, when working on the number line, we observe equivalencies:

0	1/4	1/2	3/4	1
0	0.25	0.5	0.75	1

It is important for students to see the relationship between fractions and decimals. One analogy that helps in this process is money. Consider the following table.

Money	Decimal	Fraction
50 cents	0.50	1/2 (of a dollar)
25 cents	0.25	1/4 (of a dollar)
10 cents	0.10	1/10 (of a dollar)

Number line representations also give us a way to visualize these relations.

6.10 Percentages

In fifth grade (10–11 years), we introduce the idea of percentages. A percentage is a ratio. To say that 30% of the group participated in an activity is the same as saying that a proportion equivalent to 30 out of 100 participated. It is important to relate percentages to fractions and, from there, to the part–whole idea. Later, after percentages are introduced, we integrate problems that associate them with fractions and decimals.

We can start with illustrations that will later serve as stepping stones to work with percentages.

25%	25/100 = 1/4	
33%	33/100 = 1/3	
50%	50/100 = 1/2	
75%	75/100 = 3/4	
100%	100/100 = 1	

Another idea we need to explain is that percentages, like fractions, possess an **intensive** property. Thus, the value of a percentage does not depend on its quantity or size, but on the relationship between two quantities. A useful analogy is determining the quality of jelly, which does not depend on the size of the jar, but on the proportion of fruit it contains. Here is a useful exercise in this connection.

6.10.1 *Situation*

Jellies can have different levels of quality, which often depend on the proportion of fruit to other ingredients used to make it. Higher fruit content yields better jelly.

(1) What do you think of the quality of these three types of strawberry jelly?

(2) Grape jelly is sold in large and small jars. Someone forgot the percentage of fruit content in the second jar. Can you help him? Explain your reasoning to find the percentage of fruit in the small jar.

We help students notice that, in this situation, the percentage is the same in both jars. Given that it is the same type of jelly, the size of the jar does not determine the fruit content percentage.

A percentage is a type of ratio. Thus, when saying that a jelly has a 40% fruit content, we are saying that the relation between the amount of fruit and the whole amount of jelly in the jar is a relation of 40 to 100. This ratio can be found in a small jar as well as in a large one.

The double line is a very useful analogy to discuss the concept of percentages because it reinforces the comprehension of a percentage as a fraction. For example, suppose we work with a group of 45 students. Of these, 15 are members of the Drawing Club. What percentage of the class is in that club?

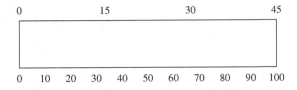

From this representation, we observe that 15 out of 45 is equal to slightly more than 30 out of 100, or a little over 30%.

References

Behr, M.J., Lesh, R., Post, T.R., *et al.* 1983. "Rational-number concept", in Lesh, R. and Landau, M (eds.), Acquisition of Mathematics Concepts and Process. New York: Academic Press.

D'Ambrosio, B. 1994. "Children's construction of fractions and their implications for classroom instruction", Journal of Research in Childhood Education, 8(2), 150–161.

Empson, S. 1995. "Research into practice: Using sharing situations to help children learn fractions", Teaching Children Mathematics, 2(2), 110–114.

Gravemeijer, K. 1994. Developing Realistic Mathematics Education. Dordrecht: Kluwer Academic Publishers.

Karplus, R., Pulos, S. and Stage, E. 1983. "Proportional reasoning of early adolescents", in Lesh, R. and Landau, M. (eds.), Acquisition of Mathematics Concepts and Process. New York: Academic Press.

Lappan, G., Fey, J.T., Fitzgerald, T.M., *et al.* 1998. Comparing and Scaling: Ratio, Proportion, and Percents. Connected Mathematics Series. Palo Alto, Calif.: Dale Seymour Publications.

Mack, N.K. 1990. "Learning fractions with understanding: Building on informal knowledge", Journal for Research in Mathematics Education, 21, 16–32.

Resnick, L.B., Nesher, P., Leonard, F., *et al.* 1989. "Conceptual bases of arithmetic errors: The case of decimal fractions", Journal for Research in Mathematics Education, 20, 8–27.

Streefland, L. (ed.). 1991. Realistic Mathematics Education in Primary School. Utrecht, the Netherlands: Freudenthal Institute.

CHAPTER 7

MEASUREMENT

7.1 Introduction

Measuring is an important topic in the study of school mathematics, given its prevalence in our everyday lives. We measure time, distance, weight, money, land, and speed, among myriads of other things. Measuring tasks are great for learning and applying the basic arithmetic operations. They also connect areas internal to mathematics as well as mathematics to other disciplines like science, art, social science, and sports.

In simple terms, measurement is a comparison of one quantity to another. When a direct comparison is not possible, an arbitrarily selected unit is used for the comparison. In more exact terms, measurement is the assignment of a numeric value to an object's attribute, like the length of a book. At more sophisticated levels, measurement consists of assigning a number to a situation's characteristic, like inflation or GDP growth.

Part of students' development of the concept of measurement is to broaden the properties that can be measured. From an early age, children are surrounded by many measuring instruments: rulers, clocks, thermometers, weight scales. Yet, it is not possible for children to completely understand a measuring system until they understand that measuring systems are indirect ways of making comparisons.

The first type of measuring tasks should be directed at helping students understand that objects have properties that can be measured. Moreover, these properties can be measured in different ways. Consider a box. We can measure its height (length), how much it can hold (volume), and how much it weighs (mass); they can also be measured using different units, be they conventional (British and metric systems) or non-conventional (hands, elbows, rocks). Let us consider length.

7.2 Measuring Length

7.2.1 *Direct comparisons (Kindergarten and First Grade/Five to Seven Years)*

Given the arguments presented in the Introduction, instead of asking children to measure the length of an object, we can present tasks in which finding the length is relevant and encourage children to focus on the attribute we want to measure. At this point, comparisons should be qualitative.

Sample activities

Who is taller?

In several tasks done in the early grades, we can ask who or which one is taller. For example, if children are playing with tower blocks, we can ask which of the towers is taller. From this task, other questions may arise that lead to a second developmental stage: indirect measurement. For instance, when comparing towers, some may be further away than others, which hampers a direct comparison. We ask students what to do and explore their suggestions as we move forward. If no child proposes using another object to compare the towers, say a piece of paper or thread (or even a ruler!), the teacher may offer a recommendation. Another idea we can tie in to this activity is keeping a record of the different heights of all the towers a child makes. That is, a child builds a tower, measures it, and stores the information in some sort of chart or list.

Another context for this activity is comparing students' heights. First, children compare themselves directly with each other to see who is taller. We can choose a wall in the classroom and assign a small section to each student where they will mark their height with masking tape and write their name on the tape.

Do the books fit?

Suppose we are arranging books on shelves. It may be that the distance between the shelves is equal, or it may not be. In either case we ask how to determine whether a books fits on a shelf. To encourage indirect measurement, we may say that we want to go to the library to check out some books to be kept in the classroom and that we want the books to fit on our bookshelf. A student may suggest measuring each shelf with a thread or paper strips (if they have different heights) and taking the measuring device to the library to measure the books.

7.2.2 *Indirect measurement (First and Second Grade/Six to Eight Years)*

As we saw in previous examples, almost any direct measurement task can lead to indirect measurement, which has several stages. In fact, this should be the process, giving students a chance to understand the reason why they are executing this type of measurement.

Global measurements

When we start taking indirect measurements, we use global measurements. For example, if we are measuring the height of a child we can write the name of the child on a piece of masking tape and put it on the wall marking the child's height, or when measuring books to see if they fit on the shelf, we can measure each book with a different string for each book.

Non-standard units

Once students have some experience with global units, we introduce a unit and "cover" the object with this unit to see how many units will fit. Initially we use different units; we call them non-standard units.

Understanding that measuring means to "cover" with a unit is an essential step to understanding the concept of measurement. For instance, when measuring the length of a table with a paper strip, we observe that we are actually covering the length with this measurement.

Normal units

In the tower blocks activity we use non-standard units, with children creating their own units. We can follow up and suggest that it might be good to gather information in some sort of chart or graph about the tallest towers built by students. We have a problem: How do we compare the measurements if each child used his or her own unit? Here the idea of using a common measurement, a normal unit, for the whole class should arise.

Suppose students suggest using a strip of paper.

We construct strips of the same size for students to measure their towers. Having a common unit facilitates making comparisons. When we have properly defined the units for indirect measurement for this task, we say we have used **normal units**. Normal units can vary from task to task, but they should all use the same unit for the same task, be it a strip, paper clips, or whatever else might be suitable. When working with normal units, it is important to observe how children are using them, checking for the following common errors:

Error 1

The starting points of measurement do not coincide. For example, if we are measuring a pencil with paper clips, we start before the beginning of the pencil:

or we start after the beginning of the pencil:

Error 2

There is space between the units:

When checking for these errors, we should emphasize that measuring means to "cover" with a unit, so it is necessary to start from the beginning of the object we want to measure, and not leave any space between the units.

Here is a sample activity to promote the concept of measurement as "covering" with units:

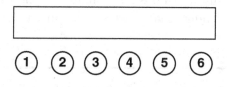

If this is the unit, ☐ , indicate how many units fit on the number line.

Once we have discussed normal units, we face situations in which the measured object does not measure an exact number of units, but one or more units plus a fraction of a unit. At this point, we should introduce the activity of calibrating the ruler. This activity helps children understand the calibration process of standard measurements, like the foot and the meter, and also reinforces the idea of fraction.

7.2.3 *Calibrating the ruler (Third and Fourth Grade/Eight to Ten Years)*

When an object cannot be measured in an exact number of units, say a paper strip, we may ask students for suggestions about what can be done. Taking their suggestions as a starting point, we can guide them until someone comes up with the idea of dividing a strip into fractions: half a strip, a quarter of a strip, and so on.[1] After calibrating the ruler, we use it in different activities to practice this new skill, while answering any questions that may arise during the process.

A potential problem in this process is the fact that length is a continuous attribute. The first tasks students work on in mathematics are related to counting. Counting requires working with discrete objects: cars, balls, flowers, etc. When we divide a discrete object, the resulting parts are no longer an example of the original object. For example, if we have a car and we divide it in two, the result is not two cars but a broken car. On the other hand, when dividing a continuous object, the divided parts continue to be examples of the original object. For example, if we divide a cup of milk in two, we still have milk in both cups. If we divide apple sauce on several plates, we still have apple sauce on each one of them.

The number line is a very important concept in mathematics, but it is more complicated than we tend to think. Children count discrete elements when they start counting. In this case, children are clear about which unit is being counted. Things change when it is time to measure, and they see this.

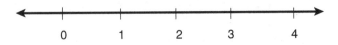

A child could ask, "What am I counting? Marks? Spaces?" Children are not clear about what defines the set that they are counting. Given this

---

[1] As discussed in the previous chapter, it is not obvious for children to realize that one-half of one-half is one-quarter. Reaching that conclusion requires time, but, as we have insisted, it is important to take the time to teach for understanding.

situation, it is common to find that when children are asked to represent 3 on the number line, they count as shown below.

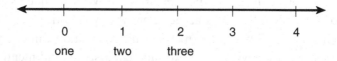

Children who do this are counting correctly, but they have decided to count the wrong set. They are counting marks or numerals, not units. However, we can use this understanding when introducing the number line by placing unit-sized paper strips above it:

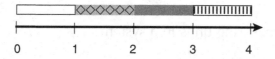

Another common error is that students start measuring from 1.

Students should be reminded that measuring is about counting the number of units that cover the length. Since from 0 to 1 there is one unit, if we start at 1 then we are counting an additional unit for the measurement of the object.

The ruler and the number line

The teaching of the number line should start from the experience of children with measurement. Developing their knowledge of the number line helps them to better understand the function of a ruler.

7.2.4 *Standard units (Third and Fourth Grade/Eight to Ten Years)*

Once students have performed measurements with a ruler that they have created in class, we introduce a situation where they need to compare their measurements with those from children in other schools. How do we know if they used the same ruler?

It is time now to introduce the standard units. We ask students to bring a ruler to class. When they look at the rulers, they will notice that there are two systems: the British and the metric. The explanation can address the historical reasons for having these two different systems, and mention that the British system is followed in England and the United States — along with those countries that have been ruled by them — but that most countries use the metric system, which is preferred in science. We can also note that the fact that two systems are widely in use points to the arbitrariness of a measurement system. However, once we fix the length of our unit, we cannot change it.

Different measuring units in a system
(fifth grade/10–11 years)

When we start measuring with standard units, with the British or the metric system, we observe that there are different units of measurement.

- British: Inches, feet, yards, miles.
- Metric: Millimeters, centimeters, meters, kilometers.

When we introduce the ruler to students, we let them exchange ideas among themselves to explore and discover the relationship between these units. The first task we must perform consists of helping students acquire an intuitive notion about the quantity of each one of these units so they can decide which one is more appropriate to measure a given item. We show different objects and ask students to identify the unit they would use to measure each of them.

We can also integrate estimating tasks into a game. For example, we show an object for students to estimate its length, then we measure the length. The student with the closest estimate to the real measurement wins.

Measuring can be integrated into multiple classroom contexts. For instance, we can perform an experiment in which we measure the growth of plants, or in sports we can measure the long jump, the triple jump, or how far a home run went. In these contexts, you can have the situation in which two students measure the same object with different units. Here we integrate learning how to convert one unit to another.

Once students are familiar with both measurement systems, we introduce an activity to measure in feet and meters. Will it be the same measurement?

Allow students to think about how to solve this situation. Discussion should lead to the idea of comparing two measurements of the same order of size, one on each system — for example, inches and centimeters.

Instead of making students remember a set of equivalencies between the British and metric systems, encourage them to remember the equivalency between a centimeter and an inch (1 in ≈ 2.54 cm), working the other equivalencies from this one. For example, we can ask them to calculate how many meters are equivalent to 6 feet. It can be analyzed in the following way:

$$6 \text{ feet} = 72 \text{ inches}$$
$$\approx 72 \times 2.54 \text{ cm}$$
$$\approx 182.88 \text{ cm}$$
$$\approx 1.8288 \text{ m.}$$

We can also show the following sketch:

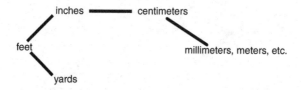

7.3 Scaled Measurements

Measurements are frequently scaled. When we read a map, we can derive distances using the scale. As with almost every concept taught in primary school, scaled measurements are not as simple as they seem and require a gradual developmental process.

Around third grade (eight to nine years), we introduce maps in which distance units are marked. Consider this picture:

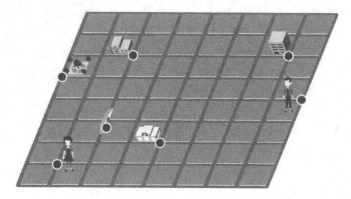

In this map, students only have to count units to determine the distances between places.

Somewhat later in the same grade, we introduce maps that mark segments representing measurement units, like in the following example:

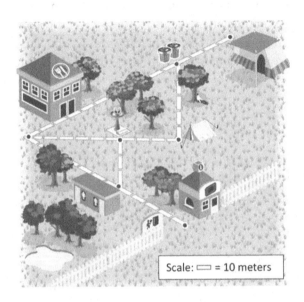

In fourth grade (nine to ten years), we use scaled maps.

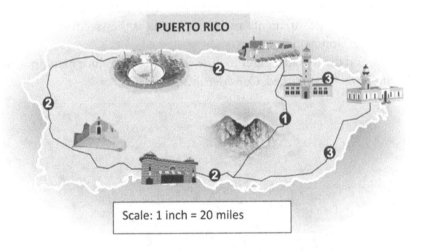

7.4 Area Measurements

7.4.1 *Idea of area (First Grade/Six to Seven Years)*

Area is not an easy concept for children to understand (Kenney and Kouba, 1997; Lindquist and Kouba, 1989). The first area-measurement tasks should be focused on helping children identify the attribute they want to measure. As stated above, objects have measurable attributes and these are measured with different units of measurement. Understanding that different attributes require different measuring units is not easy for children. Knowing which measuring unit to use, corresponding to the attribute we want to measure, is part of developing the concept of measurement. Therefore, instead of teaching children to measure area in a standard way, we must develop the concept of area of a surface; we can present tasks in which finding the area is relevant and help students identify this as the attribute that we want to measure. At this stage, comparisons should be qualitative.

Sample activities

(1) **Which space fits more children?**
    We identify two bounded spaces in the classroom and ask, "Which space fits more children?" We allow students to develop their own strategies to answer this question.

(2) **Which cookie requires more chocolate to cover it?**
    Sonia is baking some cookies for a Halloween party. She wants to cover them with chocolate. Which cookie will require more chocolate to cover it? Why?

During this problem, students may propose a strategy of direct comparison by placing one cookie over the other to see which one occupies more space. As we have witnessed, a student may also suggest measuring the circumference of each cookie with a string and then comparing the length of the strings. This is an excellent moment to discuss the relationship between perimeter and area, observing that a specific perimeter can contain different areas within it. Therefore, perimeter is not a good criterion to measure area. In fact, students should become aware during this process that the nature of what we are trying to measure calls for a type of measurement other than length.

7.4.2 *"Covering" (From Second Grade/Seven Years)*

One of the central ideas of measurement is understanding that measuring is covering the measured object with the appropriate unit. In second grade, we introduce the idea of covering space with an area measurement, and present several problems in which we use square unit measurements to measure space. For instance, consider several differently sized and shaped plots of land to plant a garden.

Plots of land

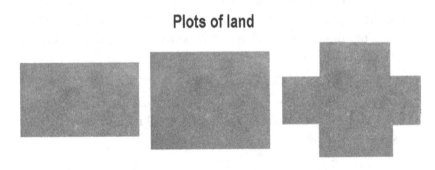

Where can you plant more carrots?

Where can you plant more carrots?

After several problems of this type, we incorporate problems, still in context, in which we place the space to be measured over a grid.

Which section of the recreational area covers more space?

Which section of the recreational area covers more space?

After students have worked on these types of problems, we provide exercises to find the areas of different rectangles, now without context but with an incomplete grid inside, like in the following problems.

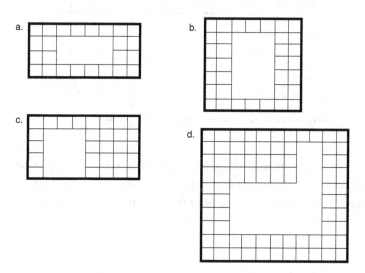

Once students have practiced this type of problem and discover that to find the total number of squares in a rectangle it is not necessary to count each of them separately — but that they can be added or multiplied — we introduce the area of the rectangle as a representation model for multiplication.

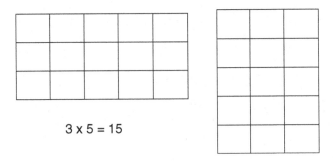

3 x 5 = 15

Redistribution

In area problems we present rectangular figures along with another type of figure, such as a triangle or parallelogram. Unlike rectangles, which have square units that "fit in" the figure, some parts of triangles or parallelograms will not fit perfectly within the grid:

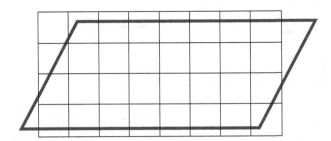

To solve these problems, we allow students to use their own strategies. For example, when working on these problems with children we have observed the following strategies to find the area of a triangle.

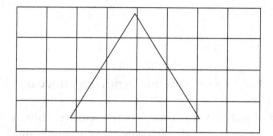

- Some students count whole units and then estimate the outlying parts (one-half, almost one-half, etc.) to add them.
- Other students observe that cutting the triangle in half through its highest vertex and transferring the one half to the other side, they can form a rectangle. Thus, the area of the triangle will be equal to the newly formed rectangle.

This second strategy integrates the idea of redistribution, a key concept in the development of area measurement. In fact, we will see later how the idea of redistribution is essential when finding formulas for the area of non-rectangular figures.

7.4.3 *Standard measurements (Fifth and Sixth Grade/10–12 Years)*

In fifth grade, we introduce square centimeters (cm²). We illustrate square centimeters, along with the centimeter and the cubic centimeter.

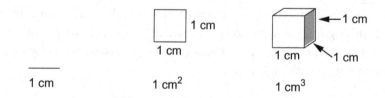

Once students have some experience with square centimeters, we explain that for each unit of length we can define a surface unit of surface.

Thus:

- In the British system we have square inches (sq in), square feet (sq ft), square yards (sq yd), square miles (sq mi).
- In the metric system we have mm², cm², dm², m², dam², hm², km².

Our main objective with these square measurements is to develop an intuitive notion of how much each square unit covers and how to decide which one is more appropriate to measure a given surface. Drawing some of them is a useful and instructive activity. Students can draw the appropriate ones in their notebooks, while others such as square feet and square meters — which do not fit in a notebook — can be traced on the floor. Then we mention different surfaces for students to decide which measurement units to use. We also ask them to estimate the areas of several surfaces, like a football stadium, a baseball park, or an Olympic-sized swimming pool.

7.4.4 *Area Formulas (Sixth Grade/11–12 Years)*

Rectangle

After working several area problems by counting and redistribution, we promote reflection. We observe that in terms of rectangles, it is not necessary to count every square unit, but that we can simply multiply the number of rows by the quantity of units in each row, or the number of columns by the number of units in each column. This is equivalent to multiplying length by width or width by length. Width gives us the number of rows and length the number of units in each row. Practicing different exercises reinforces this abstraction. For example, we can ask students to find the area of this figure that lacks explicit square units, but has some centimeter marks. They can make mental images of square units, while also associating them with length and width.

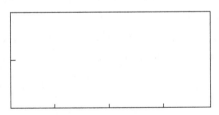

Parallelograms

A parallelogram is a quadrilateral with parallel opposite sides.

   In problems of finding the areas of figures drawn over a quadrilateral, what strategies should we use to find the area of parallelograms?

   We redistribute the figure in a way that converts the parallelogram into a rectangle with the same area.

   The area of the rectangle we have made is equal to that of the parallelogram. What is the corresponding part to the rectangle's length in the parallelogram? The base. What is the parallelogram's corresponding part to the rectangle's width? Its height. Therefore, the formula for finding the area of a parallelogram is:

> **Area of a parallelogram = base × height = b × h.**

Triangle

We start from the following activity to promote thinking that will lead to the formula for the area of a triangle.

Activity

(1) Draw any triangle. (Students will draw different triangles. Suppose they draw one like the following △ .)
(2) Make a copy of your triangle.
(3) Cut out both triangles and try to make a parallelogram with them.
(4) Observe your peers' work.
(5) Can we always make a parallelogram?

Students should be able to conclude that we can always make a parallelogram and intuitively understand that we always get a triangle when we cut a parallelogram in half along one of its diagonals. Hence:

Area of a triangle = ½ b × h.

Circle

As with the triangle, this activity invites reflection that will lead to the formula for the area of a circle.

Activity

(1) Draw a circle and, starting from its center, draw several congruent triangular regions.

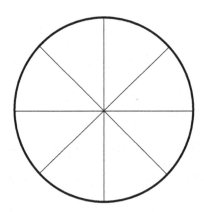

(2) Cut out the "triangles" and arrange them side by side.

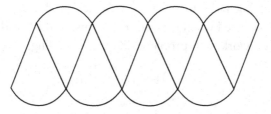

(3) What does this shape look like?

We want students to see the connection between this shape and a parallelogram. In fact, we encourage them to reflect on this and think that if we make the "triangles" very thin, the shape will look more and more like a parallelogram. We may even venture to say, without putting much emphasis on the terminology, that as the "triangles" grow thinner and thinner and their number increases to **infinity**, the **limit** will be a parallelogram.

What relationship is there between the area of this parallelogram and the area of a circle? They both have the same area!

Area of a circle = area of a parallelogram = b × h

The height of the parallelogram corresponds to the radius of the circle; the base corresponds to half of the circumference of the circle.

We can substitute those values into the formula for the area of a parallelogram.

Hence:

Area of a circle = b × h = ½ $(2\pi r)(r)$

= πr^2.

Thus,

Area of a circle = πr^2.

If we take the time to derive these formulas, students will not see them as enigmatic, but will know that they arise from the properties of the figures involved.

7.5 Volume

Before introducing tasks to measure volume, we develop the ability to visualize what we need when we want to do this. In third and fourth grade (8–10 years), we work with problems like the following.

How many boxes fit in the truck?

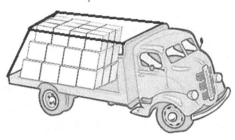

In fifth grade (10–11 years), we introduce cubic units, in particular cm³, in problems similar to the one with the truck. When students have considerable experience with problems in context, we introduce problems that call for finding the volumes of figures, without context. Eventually, students will realize that for finding the volumes of rectangular prisms it is not necessary to count all of the cubic units individually, but that they can add or multiply. We then combine volume with multiplication problems.

Find the volume.

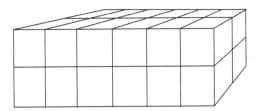

7.5.1 *Volume formulas (Sixth Grade/11–12 Years)*

The activities that we have presented to explore volume help to develop the necessary intuition to derive some basic volume formulas.

Repeated counting

At first students will count all units individually when trying to determine the number of cubic units needed to fill a recipient. Gradually, they will look for strategies to simply this process.

Development of formulas

Once students have worked on many problems like the ones described above, we ask whether they see any regularities. That should lead to the derivation of the volume formulas for rectangular prisms.

In the development of formulas for other prisms, as well as the cylinder, we want students to realize that, if they take a face and slice the figure parallel to that face, the volume will be the sum of all the layers equal to the base (that is, will be the base × height).

In the process of measuring volume, we explain that for each unit of length, we can define a unit of volume.

Thus:

- In the British system we have cubic inches (cu in), cubic feet (cu ft), cubic yards (cu yd), cubic miles (cu mi).
- In the metric system we have mm^3, cm^3, dm^3, m^3, dam^3, hm^3, km^3.

As with area, our main objective is to develop an intuitive notion of how much each cubic unit covers and how to decide which one is more appropriate to measure a given volume. Drawing or constructing some of them is a useful and instructive activity. Then we mention different volumes for students to decide which cubic units to use to measure them. We can also ask them to estimate the volume of an Olympic-sized swimming pool or an 18-wheeler cargo trailer.

7.6 Understanding that every measure is an approximation

Understanding that every measure is an approximation is difficult for students. Nevertheless, it is important to grasp this reality. Starting in third grade (8–9 years), we introduce activities that illustrate this aspect. For instance, we can ask students to measure the same object. We then compare measurements and observe the differences among them. This should lead to a discussion from which the idea of precision emerges.

In fifth grade (10–11 years), we can initiate a discussion from which the idea of a margin of errors arises. For instance, students may measure the same object and obtain the following results:

2.534 m	2.538 m	2.540 m
2.537 m	2.539 m	2.538 m
2.535 m	2.536 m	2.535 m.

We notice that all measurements are between 2.53 and 2.54 m. Students might intuitively understand that the margin of error is one centimeter.

References

Kenney, P.A. and Kouba, V.L. 1997. "What do students know about measurement?", in Kenney, P.A. and Silver, E. (eds.), Results from the Sixth Mathematics Assessment of the National Assessment of Educational Progress. Reston, Va.: National Council of Teachers of Mathematics, pp. 141–163.

Lindquist, M.M. and Kouba, V.L. 1989. "Measurement", in Lindquist, M.M. (ed.), Results from the Fourth Mathematics Assessment of the National Assessment of Educational Progress. Reston, Va.: National Council of Teachers of Mathematics, pp. 35–43.

CHAPTER 8

EXPLORING SPACE

8.1 Introduction

Until now we have addressed the development of the concept of number in children. In fact, the teaching of mathematics during the early grades is frequently limited to this topic. Nevertheless, students' spatial ability often surpasses their numerical ability.

Geometry is the mathematics of space. Mathematicians study space in a different way than do artists, architects, or even physicists; mathematicians seek a mathematical interpretation of space. The teaching of geometry should, then, focus on enabling students to develop this interpretation. In this process, consistent with our teaching philosophy, we start from the child's cognitive maturity.

Moreover, it is important to discuss elements of spatial sense that are the basis of geometric sense and which are frequently ignored during teaching. There are two basic spatial abilities: **visualization** and **orientation**. Orientation requires that subjects mentally readjust their perspective to be aware of an object's visual representation shown from different angles. However, this does not require mentally moving the object — that is the realm of visualization. These skills are essential for the geometric development of students, which must be emphasized from the early grades.

8.2 Relative Position (Pre-Kindergarten and Kindergarten/up to Six Years)

One of the first geometric tasks, while trying to organize space, is the comparison of the relative positions of two objects (Clements, 1999). We integrate activities in which students work on these concepts:

- In front of — behind.
- Above — below.
- Far — close.

We take advantage of opportunities that occur in the classroom and in the playground for students to learn these concepts (Andrews, 1996). For instance, we may warn students as they go "in front of" the swings in the playground, or ask them to "watch Nelly on the platform above the slide."

8.3 Orientation and Visualization (First to Sixth Grades/6–12 Years)

Orientation and visualization are very important for geometric intuition. Nevertheless, they are barely addressed in most curricula. From first grade, we prepare activities that develop these competences (Freudenthal, 1971). There several of aspects of this that are important.

8.3.1 *Read and orient oneself with a map*

Reading and making maps are tasks that help to conceptualize orientation and visualization (Liben and Downs, 1989). Identifying paths is an important element in the construction of these concepts. In first grade (six to seven years), we can start working on this skill by learning to identify different paths that take us from point A to point B.

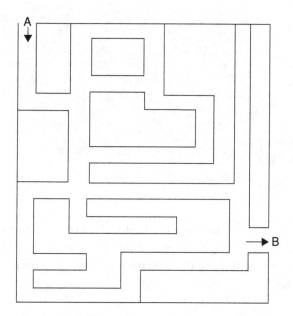

As we progress through the grades, we introduce more complex maps in which, besides paths, we integrate more points of reference.

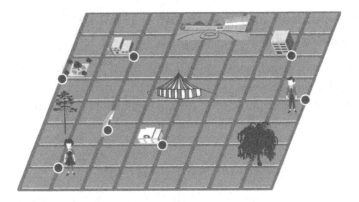

We then incorporate the idea of distance by tracing paths from one place to another and comparing different paths.

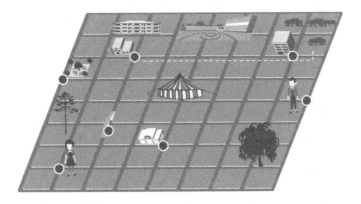

In the process of describing paths, a system of reference emerges. In second grade (seven to eight years), we put a map over squared paper. We practice two tasks: counting the squares from one place to another and a system of reference.

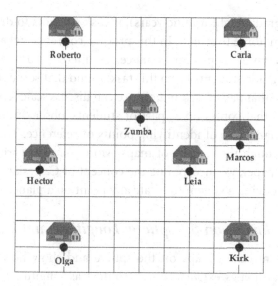

To discuss a frame of reference, we can introduce a coordinate grid and ask students to describe their paths according to it.

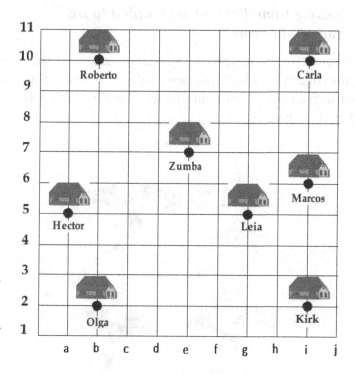

In third grade (eight to nine years), we ask students to draw the path they take from home to school. In this process, concepts will arise that are needed to interpret maps. For instance, Quintero (2001), when asking several third-grade students to do this task, found that some said, "It's not going to fit on that paper." From there, we can discuss scale representation. Another issue that comes up as part of students' observations and comments is the handiness of identifying points of reference. These skills are reinforced through the reading of maps. Moreover, when we place maps over squared paper, we incorporate the concept of Cartesian coordinates, thereby introducing a system to locate any point on a plane.

8.3.2 *Coordinates on the sphere: Longitude and latitude*

We can also introduce maps on the sphere and show how the idea of Cartesian coordinates extends to this context as longitude and latitude. This connection will help students get a deeper understanding of this way of locating places on a world globe.

8.3.3 *Looking from different angles (first to sixth grade/6–12 years)*

Another competence that promotes orientation and visualization is identifying the angle from which we're looking at an object or situation. We can start with simple activities in first grade and progressively increase the difficulty of the problems.

Where's the glass of water with respect to the vase?

8.3.4 Looking and projecting (first to sixth grade/6–12 years)

Another necessary geometric skill is projecting how a shape looks on the side we cannot see. From first grade we introduce activities in which students have to calculate how many blocks there are in a figure that they can see from a specific angle.

Can You build it? How many blocks do you need?

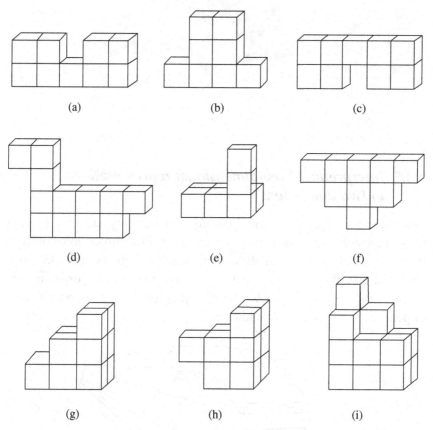

(a)	(b)	(c)
(d)	(e)	(f)
(g)	(h)	(i)

Another good exercise is asking students to analyze whether a shape can be seen from a certain angle.

What can the boy on the bench see? What can he not see?

8.3.5 *Interpreting three-dimensional representations in two dimensions*

Often we are not aware of the difficulty that students have interpreting two-dimensional pictures as representations of three-dimensional objects. It is important to work on these competences (Battista and Clements, 1995). In fact, there is much geometric knowledge used in drawing. For instance, there are several drawing techniques requiring an understanding of perspective, like an approaching train.

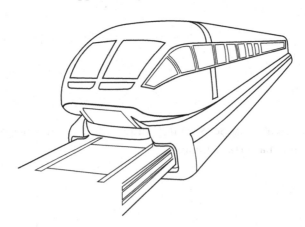

The techniques used for drawing are useful to begin a geometry lesson. Similarly, the study of perspective in geometry could be combined with a lesson on art appreciation.

8.3.6 *Geometric elements to organize space (first to sixth grade/6–12 years)*

It is important to remember that the basic geometric concepts (like point, line, and plane), as well as the basic relations between them (like parallelism, perpendicularity, and incidence) are abstractions that emerge out of a human need to describe and organize space. In elementary school, we incorporate this vocabulary in connection to some meaningful context.

As stated earlier, it is important to explore these concepts first in an intuitive way during elementary school. Then, in middle school, students will have had experiences that will allow them to reflect upon and develop — abstractly — the concepts of point, line, and plane, among many others.

8.4 Shapes and Solids

One of the most primitive mathematical tasks — in the sense of both being first explored by humans and which children do first — is observing that some objects are similar to others, from which arises the notion of similarity. Children do this naturally beginning at an early age, noticing properties that objects have in common (Clements *et al.*, 1999). Up to second grade (eight years), we should work with objects that students can touch and manipulate. For instance, while using blocks of different forms and sizes, children will notice that rectangular blocks are easier to put on top of others since all of their angles are straight, even if they do not have words to articulate those observations. Gradually, we explore more abstract properties and attributes of shapes (Burger and Shaughnessy, 1986; van Hiele, 1986).

A point of concern is that many school curricula begin working with two-dimensional figures, like triangles, squares, and circles. This is an instance of how traditional curricula follow the logic of mathematics and not the logic of learning. Although, mathematically speaking, two-dimensional figures are easier than three-dimensional ones, children first encounter three-dimensional objects. In fact, two-dimensional shapes are abstractions obtained from decomposing three-dimensional figures. To illustrate how many children might not make much sense of shapes on the plane at first, consider the following exchange.

We were once working with a group of first graders (six and seven years) and we showed them several shapes for them to classify. Immediately several interesting observations came up. The first one was that students used different criteria to classify them. The second was that the criteria used by students were not what we expected: circles, squares, and triangles. In fact, not one of them used that classification, despite being familiar with those shapes. This shows how artificial, at this initial stage, the classification of two-dimensional figures is.

The study of three-dimensional figures gives rise to two-dimensional ones. For instance, when analyzing a rectangular prism — a common toy — we observe that the faces are rectangles.

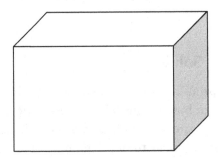

The study of plane figures, in turn, should take into consideration what we know today about learning: as with other concepts, their study should be gradual.

8.4.1 *Stage 1: Visual discrimination of shapes*

Children first identify geometric figures by their shape. Hence, we should foster visual discrimination and the identification of figures by shape, without focusing on the properties that they share and the relation between them. For instance, for a child, a square is not a rhombus because their shapes are different.

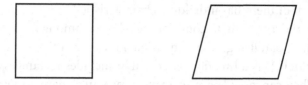

There are many activities we can integrate from the early grades that strengthen visual discrimination, like doing jigsaw puzzles. Another good activity is the following.

    a.   What elves are equal?

    b.   Pick any two elves, how are they similar? How are they different?

8.4.2 *Stage 2: Properties of objects*

For a mathematician or mathematics teacher, a square is a rhombus. They look not only at their shapes, but at the properties that define each figure. A rhombus is a parallelogram with congruent sides. Hence, a teacher calls a square a rhombus because it fits this definition; however, the student who is focusing on their shape cannot make this connection.

Research shows that conceptualization has different levels of understanding; it is a recursive process. That is, a concept is revisited, and its understanding deepens, as students develop their knowledge. A child in kindergarten can identify a square by its shape and distinguish it from a rectangle, a parallelogram, and a rhombus. At this stage, these are different concepts and none includes the other, given that their pictorial representations are different. Then, the child begins studying the properties of these shapes and seeing similarities between them. For instance:

- All have four sides.
- In all of them, all the opposite sides are parallel.

Yet, each of them has individual characteristics.

When students begin to consider the relations among these properties, they see that a parallelogram (a quadrilateral where both pairs of opposite sides are parallel) is a broader concept that includes rectangles as well as squares. They can now begin to make conceptual maps to visualize the relations among these ideas.

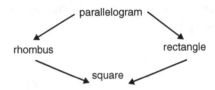

It is important to allow for the exploration and analysis of the same concept at different moments of a child's development to deepen the child's understanding. First we work at an intuitive level and, gradually, we analyze properties of the concept until we construct its definition. In this process, it is necessary to keep in mind the studies of van Hieles in 1986, who argued that learning in different areas of knowledge proceeds at different rates and therefore is at different levels in an individual. We present these levels, with geometry as the context (Burger and Shaughnessy, 1986).

Van Hiele Levels

Level 0 Visualization: Objects are the basis of knowledge. In the case of geometry, identifying geometric figures by their shape.

Level 1 Analysis: The focus of knowledge is the properties of the basic objects. For instance, observing that all sides of a square are equal.

Level 2 Abstraction: The focus of knowledge is the relationship among different properties. For example, observing that all squares are also rectangles.

Level 3 Deduction: The focus of knowledge is the sequence of sentences, like proofs, where implications and other relationships among propositions are studied.

Level 4 Rigor: The focus of knowledge is deductive systems. For this, different axiomatic systems are compared and analyzed, like Euclidean and non-Euclidean geometries.

8.4.3 *Relations (fifth and sixth grades/10–12 years)*

Once we have defined different shapes and studied some of their properties, we start to analyze the relations among the shapes and among their properties. Through this process, we elicit conjectures from students — conjectures that they will have to try to confirm and prove informally. There are several geometric software packages that can be very good in this process (Lehrer *et al.*, 1998).

8.5 Congruence

Congruence can be introduced informally in the early grades. For instance, we can show several shapes and ask students to find similarities between them. We can use the words congruence and congruent to familiarize them with this terminology, but without putting much stress on it.

Between fifth and sixth grades, we start analyzing this concept more carefully and drawing conclusions from our observations. For instance, students might notice that when two shapes are congruent, all their sides and corresponding angles are also congruent. The idea of corresponding parts requires discussion and analysis because it is not obvious for students.

Suppose we are studying congruencies in triangles. We notice that congruent triangles have congruent angles and congruent sides. From these observations, we may ask: "Do we need to confirm all of these congruencies to show that two triangles are congruent or will confirming a subset of them suffice?" A software package can be very useful in this connection.

Situations like the one described above will give students a solid intuitive understanding of the concept of congruence, which they will see more formally in high school. Moreover, the deductive reasoning skills honed through this process will give them transferable academic and life skills.

8.6 Similarity: Scale Representation

8.6.1 *Manageable representation from reality: Planes, maps, and scale models*

Previously we discussed maps as a way to reinforce orientation and localization, but a real understanding of maps requires the notion of scale.

Behind this concept, in turn, is the idea of similarity, a basic concept in many disciplines.

8.6.2 *Characteristics of two similar shapes (sixth, seventh, and eighth grades/11–14 years)*

Around sixth grade we begin the inductive study of the characteristics of two similar shapes. In this process, we discover that similar shapes have equal angles and proportionate sides. These properties have many practical applications in architecture, engineering, drawing, and photography, among other disciplines.

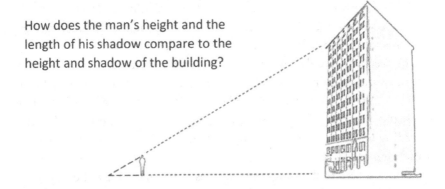

How does the man's height and the length of his shadow compare to the height and shadow of the building?

8.7 Transformations

8.7.1 *Translations, rotations, and symmetries*

Children start school with an intuitive understanding of how shapes move (Andrews, 1996). From the early grades, we can explore and observe how certain shapes slide, rotate, or move in different angles. We can also explore the concept of symmetry — for instance, when we fold a piece of paper or use a mirror. Yet, as with other concepts presented in this chapter, symmetry must be introduced gradually, studying it more deeply with each new level of presentation. In the beginning, students will have an intuitive idea of what a symmetric figure is, even without knowing the term symmetry.

As we study the properties of geometric shapes, we may show the shapes below and ask students which of them are symmetric and why that is so.

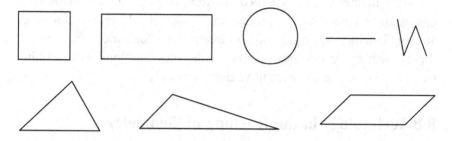

As part of the analysis of this activity, students will see that when talking about symmetric shapes, there is always a point, a line, or a plane that divides the shape into equal parts. In the case of a line, we call it an axis of symmetry.

From this concept, we introduce the shapes obtained as a result of spinning around an axis of symmetry. For instance, suppose we have square, we may ask: "How many axes of symmetry does a square have? Which figures will result from spinning a square around each one of those axes?"

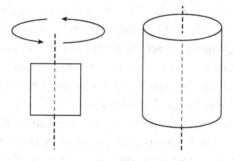

Beside the transformations caused by spinning, we discuss other types of rigid motion, like translation and reflection. In fourth grade (9–10 years), children are able to observe the effects of translation and other motions. At this point, we introduce mathematical language that helps to describe these changes. For example, a student might be able rotate a figure and now discuss the point, axis, or angle of rotation. There are software packages that let us explore these types of motions (Clements and Batista, 1989).

With mathematically talented children, one can start working with continuous transformations around sixth grade (11–12 years). Once they are comfortable with rigid and continuous transformations, we can ask them which properties they preserve under these motions. At this point, we might introduce some intuitive elements of topology.

8.8 Technology in the Teaching of Geometry

Technology can play an important role in the learning of geometry. Through the use of technology, we are able to illustrate geometric figures, their motions, and their transformations (Clements and Batista, 1989). Technology is also helpful when working with conjectures and proofs because they help us find patterns that can generate and support more conjectures. Trying to explain the behavior and existence of these patterns is a way to argue for the need of mathematical proofs.

8.9 Reasoning and Justification

Geometry has traditionally been used for the teaching of deductive reasoning in the form of mathematical proofs. However, proofs require a very gradual teaching approach. Jumping in without any informal background will certainly result in frustration for both students and teachers. Fortunately, the informal reasoning skills necessary for understanding formal proofs can be fostered in the classroom from the very beginning. For instance, every time students solve a problem or find a pattern, we can consistently ask for a justification of their reasoning — they should understand that the most important part of a problem is the creative and logical path that leads to its solution, not the solution itself. Helping students focus on their reasoning will enable them to apply their newly acquired skills to other problems or scenarios. Nevertheless, although all students should be exposed to informal reasoning, we recommend doing formal proofs only with students who have a serious interest in mathematics.

8.10 Conclusion

The study of geometry is of great value in and of itself. However, geometrical representations are also of great help for understanding other topics, like geography, drawing, physics, and architecture, among many other disciplines. Therefore, children should be exposed to different spatial situations that help them develop their visual, orientation, and motion skills. Geometrical concepts can in turn be built upon these abilities, always remembering that we construct from students' experiences and activities in context.

References

Andrews, A.G. 1996. "Developing spatial sense: A moving experience", Teaching Children Mathematics, 2, 290–293.

Battista, M.T. and Clements, D.H. 1995. 3D Geometry: Seeing Solids and Silhouettes, Grade 4. Palo Alto, Calif.: Dale Seymour Publications.

Burger, W.F. and Shaughnessy, J.M. 1986. "Characterizing the van Hiele levels of development in geometry", Journal for Research in Mathematics Education, 17, 31–48.

Clements, D.H. 1999. "Geometric and spatial thinking in young children", in Copley, J.P. (ed.), Mathematics in the Early Years. Reston, Va.: National Council of Teachers of Mathematics, pp. 66–79.

Clements, D.H. and Batista, M.T. 1989. "Learning geometric concepts in a Logo environment", Journal for Research in Mathematics Education, 20, 450–467.

Clements, D.H., Swaminathan, S., Zeitler, M.A., et al. 1999. "Young children's concepts of shape", Journal for Research in Mathematics Education, 30, 192–212.

Freudenthal, H. 1971. "Geometry between the devil and the deep sea", Educational Studies in Mathematics, 3, 413–435.

Lehrer, R., Jenkins, M. and Osana, H. 1998. "Longitudinal study of children's reasoning about space and geometry", in Lehrer, R. and Chazan, D. (eds.), Designing Learning Environments for Developing Understanding of Geometry and Space. Hillsdale, N.J.: Lawrence Erlbaum Associates, pp. 137–167.

Liben, L.S. and Downs, R.M. 1989. "Understanding maps as symbols: The development of map concepts in children", in Reese, H.W. (ed.), Advances in Child Development and Behavior, vol. 22. California: Academic Press, pp. 145–201.

Quintero, I.M. 2001. "Children map their neighborhoods", in Duckworth, E. (ed.), Tell Me More: Listening to Learners Explain. New York: Teacher College Press.

van Hiele, P.M. 1986. Structure and Insight: A Theory of Mathematics Education. New York: Academic Press.

CHAPTER 9

PROBABILITY AND STATISTICS

9.1 Introduction

We are constantly bombarded by information that is the result of quantitative reasoning. For instance, when we read the news, we may come across economic growth rates, average batting records, or life expectancy in industrialized countries. Many collective and individual decisions we make require that we understand this information and the conclusions derived therefrom. Statistics and probability help us in this process (Moore, 1990). Throughout this book, we have emphasized spiral learning. That is, during the early grades, we expose students to experiences that give them an intuitive notion of concepts, which students can cognitively access as they mature and encounter similar situations in later grades. These further encounters might include the use of symbolic mathematical language.

In this chapter we will address some basic concepts and activities in statistics and probability suitable for the early grades. Initially, we offer representations that will allow for the formalization of these concepts later on (Clement *et al.*, 1997; Konold, 1999).

9.2 Statistics

9.2.1 *Questions requiring the collection and organization of information*

Children are naturally curious and ask many questions, some of which require collecting information, like: "What are the favorite animals of each child in class? How many students have older siblings in school?"

Statistics requires that we collect and organize information in order to be able to answer our questions. We can start from the early grades and across the curriculum. In science we can ask: "How are the seedlings growing? How many days does it take a tadpole to grow legs? How does the time of sunrise change from day to day?" In social studies we may ask: "How many countries in the Americas have Spanish as the dominant language? How about English, French, or Dutch?" In English class we may ask: "How many words can you read in a minute? What is the most common letter in this paragraph?"

Knowing what data to collect to answer a question is not always a trivial issue. In fact, once we have articulated a question, we elicit from students ideas as to what type of data should be collected. Then we must decide how to organize the data. Again, we start from what students suggest, engaging them in a discussion and comparing their suggestions to help them refine their ideas.

As students familiarize themselves with the organization of information, between third and fifth grade (8–11 years) we can introduce situations requiring the study of the relations between two variables. For instance:

Do TV show preferences change with age? How about food? Sports?
Are girls' favorite TV shows different from those of boys?

The next question could come in science or social studies class:

How does temperature change with altitude? Latitude? Longitude?

Beside guiding students to study relations between two variables, we also encourage them to search for patterns and appreciate the usefulness of graphical representations in this regard.

9.2.2 *Graphical representations (starting in kindergarten/five to six years)*

One way to organize information is through graphs. We can connect graphs to topics being discussed in class across the curriculum, but remember that students should have an opportunity to read *and* make graphs. In the process, we integrate other math topics, like fractions, percentages, and operations.

Representation units

In graphs shown to children in kindergarten and first grade (five to seven years), each unit represents an object. For instance, if we are talking about the calendar, we can make a graph with the months in which children were born. On this graph, each drawing represents a child.

In second grade (seven to eight years), we introduce the idea that a unit can represent several objects:

How many CDs does each child have?

Boxes of the same size have the same number of CDs. In this case, there are 5 CDs per box.

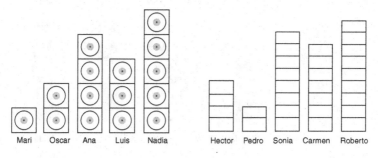

These exercises reinforce the idea of multiplication as repeated addition. Notice that we have not introduced coordinates yet, but that units are contained within the same representation.

Coordinate axes

Starting in third grade (eight to nine years), we introduce coordinate axes.

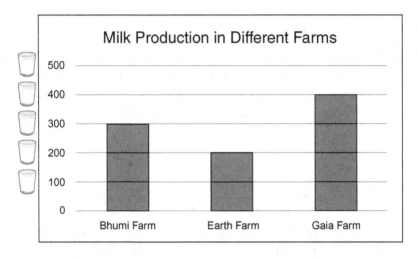

The first graphs we show, like the one above, maintain quantity representation in subdivided units. We also represent the units with cup drawings. Eventually — in that same grade — we introduce solid columns and eliminate the drawings to represent units, like in the graph below.

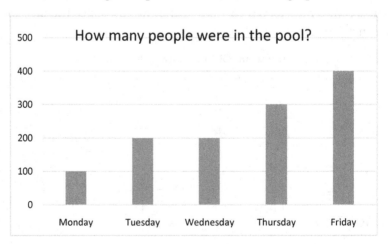

Different types of graphs

As students mature in their use and understanding of graphs, we introduce more types of graphs.

Pictographs

Picture graphs or pictographs are those in which each picture represents one or more objects. Naturally, it is more difficult for children when each picture represents more than one object. We found these mistakes while working with fourth grade students (nine to ten years):

- Some children drew figures without explaining what each represented. There were some that just drew pictures randomly, without any relationship to quantity.
- Others thought that for each category they had to draw a different picture. For instance, when representing students' favorite sports, they would draw a baseball or basketball to show those who preferred that particular sport.

After working on several examples, we introduce contexts in which the number of objects needed to be represented is large, like the number of breaths each child takes in a minute (they may draw a nose for each breath). Seeing that we would need to draw many noses, we may ask, "Is there a way to simplify this?" We are trying to elicit a response from the class, but if that does not happen, we might suggest the idea of using one picture to represent many breaths. We clarify that the number of breaths we choose that each nose will represent is arbitrary, but that once we make a decision, the number remains fixed for all instances in the graph.

Making graphs with this property serves to integrate other math topics, like counting by twos, threes, etc., or dealing with fractions when we have a quantity that is not a multiple of the fixed number chosen. In this case, you may need to fraction the drawing or choose a different number. For instance, if we have a graph in which each picture represents 4 objects and we want to represent 10 objects, we may draw 2 1/2 pictures. We may also change our arbitrary number to 2 or 5 objects per picture. This will prepare students for dealing with scales later on.

Boys in School Who Play Baseball

= 10 boys

Grade 1

Grade 2

Grade 3

Grade 4

Grade 5

Bar graphs

Understanding graph bars is a stepwise process. We first need to teach the concept of representation units and the roles of axes. Regarding the latter, we begin with graphs that only have a horizontal axis and gradually introduce a vertical axis. For instance, if we have a graph showing the height of all students in class, one axis will show height categories and the other will show the number of students per category. This distinction between the roles of axes requires special attention since it is not readily understood.

Axes

Students must realize that the decision about what each axis represents is arbitrary. In the example above, either axis could represent height, and the other the number of students.

Scales

A typical question that arises when doing graphs — even at the college level — is what scale to use on the axes. At first, students try to illustrate quantities one by one, which is fine if working with small numbers. However, they soon realize that this is not always efficient, as the following exchange we have witnessed shows.

Child: I have a problem.

Teacher: What is it?

Child: The graph doesn't fit.

Teacher: What do you mean it doesn't fit?

Child: When putting numbers on the axis, I can't get where I need to get.

Teacher: Can you think of another way of representing the numbers on that axis?

Child: I don't understand.

Teacher: Look, you have numbers going one by one. If you go by twos, then you could double the amount of numbers you can write. If you go by fives, even more.

Child: Oh! (The child leaves and begins trying different quantities until he finds an appropriate one.)

Naturally, students will show different scales to represent the same data on graphs and, within limits, different scales do the job well. However, beside clarifying that scales are arbitrary, we must also stress that there are scales more appropriate than others — and teach them how to choose the right scale. For instance, a very small scale might make it very difficult to represent large numbers; a very large one might distort the idea we want to convey. In fact, deceptive advertising makes use of this distortion. In sixth or seventh grade (11–13 years), we can begin to teach how to identify this type of deceptive graph.

Sample activities

Making graphs is an activity that allows us to easily integrate different topics of interest for students. For instance, we can graph the Major League Basball batting leaders or the time certain songs have remained as the leading single in the *Billboard* charts.

Line graphs (fourth grade/nine to ten years)

We integrate line graphs in fourth grade. These can be seen as extensions of bar graphs, as shown below.

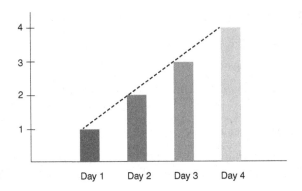

Plant Growth during First Week after Sprouting

As with bar graphs, line graphs show a relation between two variables. However, in line graphs, the variable is continuous instead of discrete (and, hence, represented by a line). We identify several points and join them with a line. Each variable is represented on a different axis. As illustrated below, line graphs become more and more abstract.

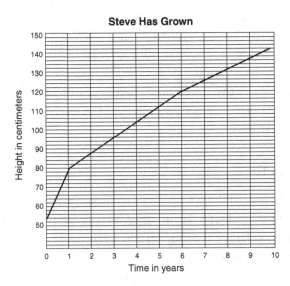

Steve Has Grown

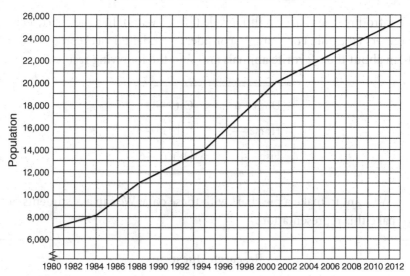

Population Growth in the Town of San Isidro

Multiple-bar graphs (fifth and sixth grades/10–12 years)

As information becomes more complex, represented categories may have sub-categories. Multiple-bar graphs allow us to show this.

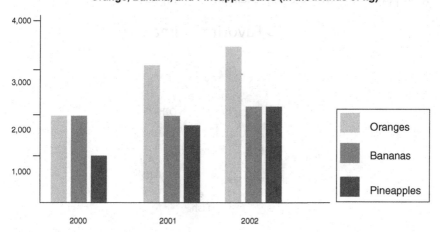

Orange, Banana, and Pineapple Sales (in thousands of kg)

Pie charts

We introduce pie charts in fifth grade (10–11 years). In a pie chart, a circle represents the unit. For example, consider 24 students in a classroom. We ask them about their favorite animals and find the following results.

Dogs—6	Cats—4
Horses—3	Rabbits—1
Fish—4	Others—6

We want to make a pie chart with these data. We can ask students how to make this representation.

Research Questions

(1) Are students capable, at this stage, of relating this problem to fractions?
(2) If so, what interpretation of fractions should students use to solve this problem?

Once students have exchanged some ideas, we discuss them and introduce fractions. The first task is to find the number of students and see what part of the whole group each category represents.

Favorite Animals

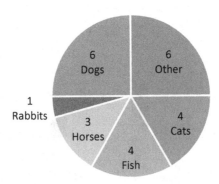

Pie charts are good for deepening the concept of percentages. Once students understand the concept, we explain that the entire pie represents the total unit — that is, 100%. For the problem above, we get these percentages:

Total = 24 → 100%
Dogs = 6 → 6/24 = 25%
Horses = 3 → 3/24 = 12.5%
Fish = 4 → 4/24 ≈ 16.7%
Cats = 4 → 4/24 ≈ 16.7%
Rabbits = 1 → 1/24 ≈ 4.17%
Other = 6 → 6/24 = 25%.

Favorite Animals

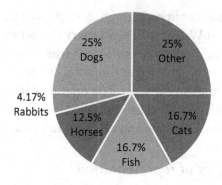

While analyzing this graph, we notice the equivalence between fractions and percentages, and reflect upon how convenient it is to learn some of these equivalences, like:

50% = 1/2,
25% = 1/4,
75% = 3/4,
33.3% = 1/3.

We also practice rounding off numbers and explain the difference between an equivalence and an approximation. In addition, we can connect this concept to money: $1 = 100% = a whole, 25¢ = 1/4, and so on.

Comparing graph information (sixth grade/11–12 years)

Sometimes we have different graphs about the same issue. For instance, the following graphs show the results of a math test at two different schools.

What problems do we have when comparing the information on these graphs?

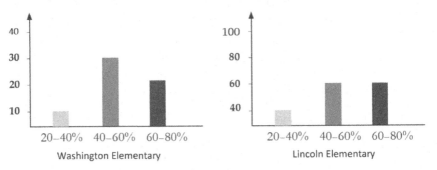

	Washington Elementary			Lincoln Elementary	

The scale is different in each graph. At first glance, we get the impression that students at Washington Elementary did better, but upon careful examination, we realize that the opposite is true. That is why it is important to choose the same scale when comparing graphs.

We can also represent the same data on different graphs, like in a bar graph, a line graph, and a pie chart. As a good exercise to practice this skill, we can give students some data and ask them to produce different graphs.

9.2.3 *Other types of representation*

Oftentimes, we have to group information by categories. For instance, the Perez Family is calculating how much it will cost to go an amusement park. These are the fares, along with the family members and their ages (in years).

Perez Family Members and Their Ages

Mari — 38	Luis — 15
Rafael — 37	Ricardo — 40
Melinda — 10	Rafi — 10
Lisa — 18	Ana — 35
Ileana — 8	Erick — 6
Fernando — 44	Javier — 14
Don Paco — 67	Nilsa — 42
Carlos — 12	Doña Laura — 64
Hector — 17	Alexandra — 9

Fares by Age

0–4	Free
5–12	$8.00
13–24	$12.00
25–59	$10.00
60 and up	Free

To calculate the total, we tabulate the data to see how many members fall into each of the different fare categories and then multiply those numbers by the category price to conclude that the total is $156.

Age	Fare	Number of Members	Price
0–4	Free	0	$0
5–12	$8.00	6	$48
13–24	$12.00	4	$48
25–59	$10.00	6	$60
60 and up	Free	2	$0

We can introduce similar activities where tabulating data is convenient. However, we should note that when we do this, we sometimes lose some information we may need later. For example, consider a teacher who is looking a test's grade distribution. She uses the following scale.

100–90	A
89–80	B
79–70	C
69–60	D
59–0	F

Suppose there were three A's. Using a chart similar to the one about the amusement park, she will not know, by looking at the chart, how high or low were those A's. A graph that helps in this type of situation is a stem-and-leaf plot.

Stem-and-leaf plots (fifth or sixth grade/10–12 years)

This type of graph allows us to represent the grade distribution without losing pertinent information. For example, suppose the teacher above gives a test to her 32 students and they get the following grades:

70	80	93	35	53	71	87	59
62	70	79	77	65	71	95	84
62	54	74	63	65	71	89	78
87	74	73	68	79	98	83	46

We must first decide how we want to subdivide the data. In this case, we will have two columns: the left one for the tens (called the **stem**) and the right one for the ones (called the **leaf**).

Stem	Leaf
0	
1	
2	
3	5
4	6
5	3 4 9
6	2 2 3 5 5 8
7	0 0 1 1 1 3 4 4 7 8 9 9
8	0 3 4 7 7 9
9	3 5 8

Given how the data are represented, we can quickly recover the original information.

9.2.4 *Describing data*

Summarizing data

From kindergarten to second grade (five to eight years), children look at facts individually. They might know that there are three children in their class whose birthday is in October, but they do not look at the data as a whole. Around third grade (eight to nine years), we can begin promoting that students look at data collectively, looking for patterns and trends (McClain, 1999). For instance, if we have data about modes of transport to school, we want them to observe that most students come, say, by car in a suburban area, by subway in some New York City schools, or by bus in some rural areas. Statistics offers measures to describe trends in data, such as the **average** (mean, median, and mode) and **dispersion** (variance and standard deviation).

The most well-known way of describing data with a numerical value is the average — specifically, the arithmetic mean. As we have repeatedly stated, it is important that formulas be introduced in context and not just by themselves as if acts of magic. In fourth grade (nine to ten years), we can introduce the concept of average (Mokros and Russell, 1995). We show some activities to help reinforce the teaching of this concept.

Activities

Average

From exercises such as these, the arithmetic mean formula arises.

(1) After breaking a piñata, John picked 12 candies, Luis 9, Mary 8, Laura 5, and Rick 1. How many candies would each have picked if everyone had got the same amount?

(2) Mike scored 20 points in the first basketball game, 12 in the second, 32 in the third, and 8 in the fourth. Gabriel scored the same total of points, but scored the same amount in each game. What was Gabriel's score per game?

(3) Ana spent 35¢ for lunch on Monday, 22¢ on Tuesday, 33¢ on Wednesday, 10¢ on Thursday, and 20¢ on Friday. Next week, she will have the same amount of money but she wants to spend the same amount each day. How much will Ana spend every day?

However, we note that although many people might know the arithmetic mean formula — thinking of it as *the* average formula — few fully understand the concept (Uccellini, 1996). Many think that the arithmetic mean is:

(1) The value that occurs most often (the mode).
(2) The value in the middle if we arrange the data in ascending or descending order (the median).
(3) The value closer to most data.

Although the mean might have these attributes, it might not. Let us see some examples.

Example 1: Are humans taller today than yesterday?

We begin this activity by asking, "Are humans taller today than yesterday? How can we tell?" After eliciting several ideas from students, we may suggest — if not already offered by them — to do a study on how height has changed from father to son and mother to daughter, in case there is a difference due to gender. We may ask students to collect information from their family members and neighbors over 20 years old. Moreover, we explain why it is more convenient to collect the information in centimeters rather than feet and inches; alternatively, they may change feet and inches to decimal notation (e.g. 5.8″ to 5.67′). Once the information is gathered, we can group students and ask them how to organize the data. Then each group can explain to the class the method they chose. Collectively, teacher and students choose one method (or propose an alternate one) and the teacher demonstrates it on the board.

Suppose we select the male information first, measured in centimeters.

Son	Father
172	174
199	189
170	173
192	178
195	185
191	194
185	180
180	184
180	166
164	167

We explain that to calculate the average (mean) height of the fathers (179 cm), we add their heights and divide by the number of fathers (10 in this case); we repeat the same process with the sons' data.

After doing this, a student is surprised that, although she calculated the fathers' average height correctly, no individual father had that height. You may ask the student why she is surprised and then ask the group: "Is anyone not surprised? Why are you or not surprised?"

This example shows that the mean is not necessarily the most common value — in fact, it might not even be a value!

Let us look at an example that shows that the mean is not necessarily the middle value, for data organized in an ascending or descending order.

Example 2: Wages

In a very small country, a proud president announces that the average per capita annual income is $58,400. However, most people complain about low wages. What could be wrong?

After collecting and organizing the information, we find the following wages per inhabitant:

Wages	Number of people
$450,000	1
$150,000	1
$100,000	2
$32,000	4
$20,000	12

Notice that in this case, the average (mean) salary does not reflect what the majority of the people earn; indeed, most people from this country have wages that are lower than $58,400. If we graph the data on the number line, we will see that most wages fall to one side of the graph, but that the few very high wages tilt the mean away from them. We describe this situation by saying that the average (mean) is very susceptible to extremes.

Example 3: Median and mode (sixth and seventh grades/11–13 years)

In the problem above, the average (mean) salary was not in the middle of the data. If we arrange the wages in ascending order, which salary is exactly in the middle? Since there are 20 facts, we get two values in the middle, but both are $20,000. Although that is not the mean, it is a type of average we call the **median**. Let us consider a few examples.

We want to find the median of the following data:

$$23, 25, 14, 56, 43, 29, 32$$

We first arrange the numbers in ascending or descending order. Thus:

$$14, 23, 25, 29, 32, 43, 56$$

Since 29 is in the middle, that is the median.

The **median** is a measure of central tendency. In some cases, the mean is a better representation of central tendency than the median, while in other situations, this is the other way around. However, it is better to consider both to have a clearer idea of the central tendency of the data.

In the wages problem the median coincides with the most common value, called the **mode**. This is not always the case. For instance,

$$38, 42, 50, 54, 55, 55$$

As suggested in the wages problem, if the number of individual data is even, then we do not have a middle value. Hence, to calculate the median, we take the mean of the two values in the middle. Therefore, the median in this case is $(50 + 54)/2 = 52$, whereas the mode is 55 and the mean is 49.

The mean, median, and mode are all measures of central tendency. The mean tells us the weighted average; that is, if we take all the values and redistribute them equally. The median tells us the value that lies in the middle, even if that particular value is not a member of the data set. The mode tells us the most common value. To get a better picture of the central tendency of a data set, it is good to know all three measures.

Example 4: Dispersion measures (sixth grade/11–12 years)

Although mean, median, and mode offer properties of the distribution, they do not tell us how spread apart the data are. This property is called **dispersion**. Let us see an example.

There are two small towns with 10 inhabitants each. These are their salaries:

Town 1	Town 2
6200	1000
6000	2000
6700	3000
6900	4000
7400	7000
7600	8000
7900	9000
8000	11000
9100	14000
9200	16000

A student says that the conditions are similar in both towns because the salaries have the same mean and median (and mode, since there is none). Do you agree with her? Why?

This situation shows the need of defining a measure in addition to those of central tendency. This is a case where the measures of central tendency coincide, yet circumstances in both towns are very different. How are they different? Let us graph the data on a number line.

Town 1

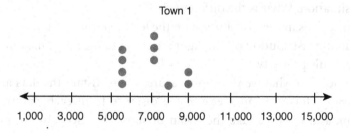

Town 2

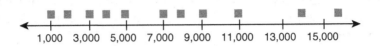

What do you see?

In Town 1, wages are close to the mean, whereas in Town 2, the values are more spread out. We need a measure that will reflect how spread out, or how dispersed, the data are. In other words, we need to find measures of dispersion. How can we do this?

There are two characteristics of dispersion that we can measure. A simple one is how far the smallest datum is from the largest. For instance, in the problem above we have:

$$Town\ 1: 9200 - 6200 = 3000,$$
$$Town\ 2: 16000 - 1000 = 15000.$$

We call this measure the **amplitude**. The amplitude only tells a partial picture of dispersion. Consider the following two graphs.

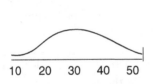
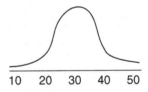

Both distributions have the same mean, median, and mode: 30. They also have the same amplitude: 40. Nevertheless, they do not represent the same situation. What is the difference?

The values in the distribution on the left are more spread out, whereas those in the distribution on the right are closer to the median. How can we measure this property?

Notice that what we want to measure is how distant the data are from the mean. Thus, we can measure the distance from each datum to the median, and then we take the mean of those distances. In the problem above, the median is $7,500.

	Town 1		Town 2
Salary	Distance to the mean	Salary	Distance to the mean
6200	1300	1000	6500
6000	1500	2000	5500
6700	800	3000	4500
6900	600	4000	3500
7400	100	7000	500
7600	100	8000	500
7900	400	9000	1500
8000	500	11000	3500
9100	1600	14000	6500
9200	1700	16000	8500

Calculate the mean of the distances in each group.

$$\text{Group } 1 = 8600/10 = 860,$$
$$\text{Group } 2 = 41000/10 = 4100.$$

Notice that in Group 2, where the data are more dispersed, the mean of the distances to the median is greater. This is one way of measuring dispersion.

The measure most commonly used to determine dispersion is the **standard deviation**. This measure is close to the measure above, but is different. We can, however, think of it as measuring the average of the dispersion of the data to the mean. Finding the standard deviation is a task we would leave for seventh or eighth grade (11–14 years).

9.2.5 *Representativeness (sixth grade/11–12 years)*

One of the fundamental ideas in statistics is the representativeness of data. How can we, with the given information, draw conclusions that may apply to the entire population? This is one of the most difficult questions in statistics (Schwartz *et al.*, 1998). For this we need a sample, since it is usually impossible or impractical to account for every member of a population. A **sample** is a subset of the population, which, given its characteristics, is thought to represent the entire population. For example, when trying to determine the voting tendencies of an electorate, it is necessary to take a sample. For the sample to be representative of the population, it must contain people of different ages, genders, geographical locations, and socio-economic status, among other factors. Each of these groups must be represented in the same proportion as in the entire population. After this explanation, students will come to realize that taking a good sample is not a simple task. Nevertheless, we can expose primary school students to this idea.

Activity: Are humans taller today than yesterday? (Part 2)

In the problem about the height of parents and children, we used the information provided by students. Can we draw valid conclusions for the entire population of our country? What about for the world's population? For our conclusion to be valid, the sample must be representative. Consider the following example.

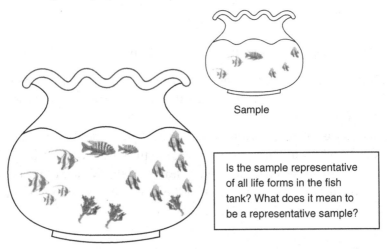

Sample

Is the sample representative of all life forms in the fish tank? What does it mean to be a representative sample?

We can work with similar examples to construct the idea of representativeness.

9.3 Probability

Patterns and relations studied in mathematics arise primarily from the observation of nature. Initially, humans studied patterns that occurred with a level of certainty. For instance, when tossing a ball in the air, we know with certainty that it will fall. These regularities, like the speed of a falling object, were expressed in terms of the laws of physics. However, when spinning a roulette wheel, we cannot be certain whether the arrow will land on a red or a black spot, much less of the exact number on which it will land. This is an example of an uncertain outcome.

Around the 18th century, mathematicians began to tackle this problem of uncertainty, creating a branch of mathematics known as **probability** — a model of uncertainty. It is important that students understand that probability attempts to quantify the levels of certainty of different events (Konold, 1989).

From kindergarten to second grade (five to eight years), we can informally introduce ideas about the uncertainty of a situation (Konold, 1999). Teachers may use vocabulary that allows them to talk about uncertain situations, like "We will probably go to the playground this afternoon," or "It is probable that it will rain today." In third grade (eight to nine years), we begin to generate the idea that there are phenomena that are more probable than others with activities like the following.

9.3.1 *Activities*

Activity 1: Will it happen? (third or fourth grade/8–10 years)

In the following table, indicate if you are certain that the given situation will occur, or if you are certain it will not occur.

Event	I'm certain it will happen	I'm not certain it will happen	I'm certain it will not happen
It will rain tomorrow			
A dinosaur will come to class			
My mother will come to school			
If I toss a coin in the air, I will get heads			
If I throw a ball in the air, it will eventually fall to the ground			
My team will win the game			

Activity 2: How certain am i? (fourth grade/9–10 years)

Mark on the line to the right how certain you are that the given event will occur.

I'm very certain that it will happen. →

1. I will go to school tomorrow.
2. If I roll a die, a 2 will show.
3. If I toss a coin, I will get heads.
4. If I throw a quarter in the water, it will sink.
5. Tomorrow I will be able to fly.

I'm very certain that it will not happen. →

We observe that we can order the probability of an event from more to less certain. Probability will quantify the likelihood of a given outcome.

Activity 3: What's the probability? (fourth and fifth grades/9–11 years)

Indicate the probability that each of these situations occurs. If you are completely certain it will happen, the probability will be 100%, or 1. If you are certain it will not occur, the probability will be 0%, or 0.

First floor   Second floor

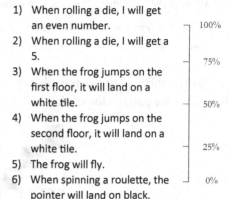

1) When rolling a die, I will get an even number.
2) When rolling a die, I will get a 5.
3) When the frog jumps on the first floor, it will land on a white tile.
4) When the frog jumps on the second floor, it will land on a white tile.
5) The frog will fly.
6) When spinning a roulette, the pointer will land on black.

100%

75%

50%

25%

0%

Activity 4: Who will go? (fourth and fifth grades/9–11 years)

A travel agency gave away one ticket to Disneyworld to a graduating class. The teacher wrote the names of students on separate pieces of paper and put them in a box to choose the winner by drawing one at random. Do you think this is a fair method?

The graduating class has 30 students:

- 15 boys and 15 girls.
- 20 are from an urban area and 10 from a rural area.
- 5 students are older than 12 years of age.

(1) What is the probability that the teacher will choose a boy? What about a girl?
(2) What is the probability that the winner comes from a rural area? What about from an urban area?
(3) What is the probability that the winner will be older than 12?

Activity 5: A frog in the hallway (fifth and sixth grades/10–12 years)

Esther has a pet frog. When Esther went to the cafeteria, the frog jumped onto the floor. The cafeteria's floor has the following pattern:

What is the probability that the frog landed on a black tile?

Draw different patterns where the frog will have the same probability of landing on a black tile as it does in the cafeteria.

Esther then went to the library and the frog jumped again. The library's floor pattern looks like this:

What is the probability that the frog landed on a black tile?

Draw different patterns where the frog will have the same probability of landing on a black tile as it does in the library.

Where is it more probable that the frog will land on a black tile, in the cafeteria or in the library? Why?

From these activities, students might conclude that the probability of an event is related to the number of possible outcomes and the number of outcomes in which the event occurs. We can tell them that we refer to this as the **probability** of an event A, or $P(A)$, and express that relationship in the following formula (or, better yet, try to elicit the formula from students).

$$P(A) = \frac{total\ number\ of\ outcomes\ where\ A\ occurs}{total\ number\ of\ outcomes}.$$

This formula can be introduced in sixth or seventh grade (11–13 years). Activities such as the ones in this section should support its interpretation.

9.4 Concluding Remarks

Statistics helps us to decide which data to collect, how to organize those data, and which graphical representations support its understanding. While studying many curricular topics, questions arise whose answers require the collection, organization, and interpretation of data. These questions can be the basis on which to build the study of statistics.

The study of statistics offers tools for a critical analysis of the quantitative information that students constantly receive through the media. This information serves as learning material. The development of skills for statistical reasoning contributes to the education of informed citizens who will be able to make decisions based on the evidence and by applying the appropriate quantitative reasoning.

While statistical inference is studied at the secondary level, we can informally initiate students into the study of these concepts in elementary school. Hence, the students' understanding can mature gradually.

As we have emphasized throughout the text, the process of teaching these topics should follow the logic of learning, building from the knowledge students possess. Formalizing these concepts and their symbolic representations should occur once students have developed an intuitive understanding.

References

Clement, J., DiPerna, E., Gavin, J., *et al.* 1997. Children's Work with Data. Madison, Wis.: Wisconsin Center for Education Research, University of Wisconsin.

Konold, C. 1989. "Informal conceptions of probability", Cognition and Instruction, 6(1), 59–98.

Konold, C. 1999. "Probability, statistics, and data analysis", in Kilpatrick, J., Martin, W.G. and Schifter, D. (eds.), A Research Companion to NCTM's Standards. Reston, Va.: National Council of Teachers of Mathematics.

McClain, K. 1999. "Reflecting on students' understanding of data", Mathematics Teaching in the Middle School, 4(6), 374–380.

Mokros, J. and Russell, S. 1995. "Children's concepts of average and representativeness", Journal for Research in Mathematics Education, 26, 20–39.

Moore, D.S. 1990. "Uncertainty", in Steen, L.A. (ed.), On the Shoulders of Giants: New Approaches to Numeracy. Washington, D.C.: National Academy Press, pp. 95–137.

Schwartz, D.L., Goldman, S.R., Vye, N.J., *et al.* 1998. "Aligning everyday and mathematical reasoning: The case of sampling assumptions", in Lajoie, S.P. (ed.), Reflections on Statistics: Learning, Teaching, and Assessment in Grades K-12. Mahwah, N. J.: Lawrence Erlbaum Associates, pp. 233–273.

Uccellini, J.C. 1996. "Teaching the mean meaningfully", Mathematics Teaching in the Middle School, 2(2), 112–115.

CHAPTER 10

PATTERNS, RELATIONS, AND FUNCTIONS

10.1 Introduction

It has been said that mathematics is the study of patterns and relations (Steen, 1990). In fact, patterns and relations observed in nature give rise to mathematical concepts. In turn, mathematical concepts and relations become the object of analysis from which new patterns, relations, and mathematical structures emerge. To understand the nature of mathematics, it is essential to experience the discovery and expression of patterns and relations.

Given the important role patterns play in mathematics, learning to find them is necessary for their understanding. For example, when learning the numbers 1–10, it is important to memorize the sequence 1, 2, 3, 4, 5, 6, 7, 8, 9, 10. However, since there are infinitely many positive integers, it is impossible to memorize all of these numbers. Yet, if we focus on the pattern in the tens:

$$10, 11, 12, 13, \ldots, 19$$
$$20, 21, 22, 23, \ldots, 29$$
$$30, 31, 32, 33, \ldots, 39$$

then we only need to learn numbers 1–10 and then apply the pattern. Naturally, the tens follow the same pattern: 10, 20, 30, 40, 50, 60, 70, 80, 90.

If we can identify these two patterns, then we can quickly learn numbers 1–100. Since the other natural numbers are constructed in the same way, we can potentially keep counting *ad infinitum*, even if we do not have names for those numbers.

The teaching of mathematics must integrate activities of finding patterns and relations, as well as promoting that students express those patterns in mathematical language. These activities can be both in the form of group discussion and through exercises and problems devoted to that end. *In fact, it might be advisable to begin the teaching of mathematics with the study of patterns, even prior to the introduction of numbers.*

10.2 Notion of a Pattern

Searching for patterns and relations is something we do from an early age. As infants, we intuitively discover gravity by tossing toys in the air and watching them fall. After this discovery, we can predict that if we toss an object in the air, it will fall. In a similar manner throughout history, humans have been identifying patterns and relations that help them to understand the world and predict future events.

We can use examples from students' experience to explain this concept. For instance, we can use nature's patterns, like night–day–night and rain–rainbow, or patterns that occur in students' daily lives, like clean–dirty–clean and hungry–eat–hungry. Gradually, we introduce problems that develop their ability to identify patterns or, better yet, help them to nurture their natural taste for *seeking* patterns. This might be advisable to do even prior to teaching them the number sequence, since not only is this a more basic skill than counting but, as shown above, it is also the basis for counting. Consider the following introductory problems for young children.

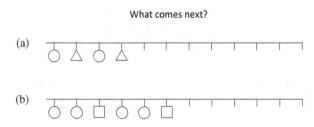

What comes next?

(a)

(b)

10.3 Patterns and Relations During Conceptual Development

Mathematics is ripe with patterns. Hence, comprehension of many concepts is obtained or strengthened by finding patterns. Consider the following examples.

10.3.1 *Amount and one-to-one correspondence*

In Chapter 3, we discussed how the idea of number is based on the idea of a one-to-one correspondence. Often, establishing a one-to-one correspondence is attained by observing patterns. Consider the following example.

Are there more black or white dots? Are there the same number?

One way to solve this problem is just counting. However, if students notice a pattern like ⬚○● and realize that it is equivalent to ⬚●○, then they might be able to solve the problem quickly by going down six-dot columns ●○ and realizing that the last (three-dot) column has two white dots but only one black dot. Hence, there is an extra white dot. Notice that we solved the problem without knowing exactly how many dots of each color there were. This is an instance of economy of thought, an important trait for identifying mathematical talent.

10.3.2 *Counting and the learning of numerical symbols*

When children learn to count and write numbers, we should foster the discovery of patterns in our number system that will help them to remember numbers and, later, discover some of their properties.

As mentioned in the introduction to this chapter, being aware of some basic patterns of our number system allows us to make important connections. In fact, we witnessed the following exchange in a second-grade class (seven to eight years) in which a girl was counting "57, 58, 59," and then stopped not knowing what number came next.

Teacher: With what number does 59 start?

Girl: With five.

Teacher: And what number comes after five?

Girl: Six.

Teacher: So, what number comes after 59?
Girl: Sixty!

Thinking of the 1, 2, 3, ... sequence and seeing the relation between that and the numbers 10, 20, 30, ... allowed the girl to continue counting.

10.3.3 *Before-and-after notion*

Many students have difficulty with the "before" and "after" notions, like in the following example.

> Fill in the blanks with the missing numbers.
> ____, 48, 49, ____
> 36, ____, 38, ____
> ____, 70, ____, 80

Recognizing patterns in our number system helps in this type of exercise. When we teach the concepts of "before" and "after" we should concentrate on explaining this relation between numbers 1 and 10. Once this is clear, we guide students to see the connection to the tens. Thus, since 7 < 8, then 37 < 38, 97 < 98, etc. Likewise, 20s are less than 30s, 30s less than 40s, and so on.

10.3.4 *Relations in the number sequence*

It is obvious to us that the difference between two consecutive natural numbers is one. However, this is not readily so for students when they first encounter numbers; once they understand this relation, it reinforces and supports them in doing calculations. Through representations of numbers and constantly working with them, we help students see this relation.

10.3.5 *Arithmetic operations*

Patterns and relations support the learning of the basic combinations of addition, subtraction, multiplication, and division.

Addition

When learning the basic addition combination, it is natural to remember some better than others. Those we remember can serve as a foundation to construct other combinations, like if we see the one-more/one-less relation. Thus, if we remember that

$$4 + 3 = 7$$

and I must find

$$5 + 3 = \underline{}$$

we note that 5 is one more than 4. Hence, $5 + 3 = 8$ (one more than 7). Another pattern we must reinforce is adding 10.

$$10 + 5 = 15,$$
$$10 + 7 = 17,$$
$$10 + 8 = 18.$$

Multiplication

While working on the arithmetic operations, we should always encourage the searching for patterns and relations. Consider the following example:

$$\text{If } 25 \times 4 = 100,$$
$$\text{then } 25 \times 40 = 1000,$$
$$\text{and } 25 \times 400 = 10000.$$

Also, if we observe that:

$$15 \times 16 = 240$$

then 14×16 must be $240 - 16 = 224$, since 14 is one less than 15.

Thus, exercises promote practice as well as the searching for patterns.

10.3.6 *Other concepts*

As stated earlier, the teaching of mathematics requires the identification of patterns and relations. This makes the learning of mathematics more appealing and helps students to remember with understanding.

10.4 Patterns and Relations at the Heart of Algebra

During the study of algebra, students work primarily with variables as number representatives, as well as with equations and inequalities, which condition those variables. The teaching revolves around methods to transform those equations or inequalities to obtain the value or values that make them true. This offers a very limited view of algebra. In fact, the heart of the study of algebra is the study of patterns and relations among different mathematical entities (Fey, 1990). Often, these relations and patterns are expressed as equations; some of the relations are functions.

The National Center for Research in Mathematical Sciences Education at the University of Wisconsin, following the lead of the Freudenthal Institute, created the Mathematics in Context series, in which they develop the concepts of algebra from the finding of patterns and relations. We show several examples from that series that demonstrate how we can introduce the idea of a variable as a tool to express patterns.

Maribel is fixing her yard. She wants to build a patio. To do this, she will put tiles on that part of the yard. She is planning to use two types of tiles and begins by drawing patterns of different lengths. These are the first patterns she thought of:

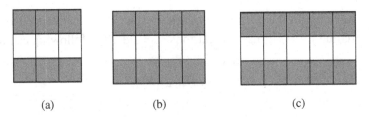

(a)　　　　　　(b)　　　　　　(c)

What relation is there between the number of tiles and the pattern length?
Can you express this relation using variables?

Let n be the pattern length (number of columns). Then, the number of tiles in A, B, and C will be 3n.

> *What is the relation between white tiles and pattern length? What about between the black tiles and the pattern length?*

The white tiles will be n and the black ones 2n.

Maribel thought the pattern above was too boring, so she decided to change it. She took pattern A and added two new columns.

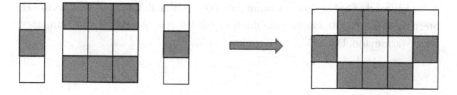

> *How would the new pattern look with B and C?*

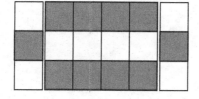

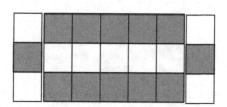

> *What relation is there now between pattern length and the number of tiles?*

The white ones are n + 4 and the black ones are 2n + 2.

10.5 Conclusion

We reiterate that being able to identify patterns and relations is essential throughout the teaching of mathematics — starting from the early grades — since finding patterns helps us to better understand mathematical concepts and methods.

References

Fey, J.T. 1990. "Quantity", in Steen, L.A. (ed.), On the Shoulders of Giants: New Approaches to Numeracy. Washington, D.C.: National Academy Press, pp. 61–94.

Steen, L.A. (ed.) 1990. On the Shoulders of Giants: New Approaches to Numeracy. Washington, D.C.: National Academy Press.

CHAPTER 11

THE JOY OF PUZZLES

11.1 Introduction

When selecting problems for class, we should remember that the "realistic" contexts students bring to the classroom involve aliens flying across galaxies; wizards, elves, and gnomes inhabiting enchanted forests; dragon-fighting knights rescuing princesses from mystical castles; superheroes saving the world; and zombies, vampires, and evil creatures haunting our cities, among a myriad of fantastic creatures and scenarios. We should use these elements and contexts to create problems that capture students' attention and feed their imagination.

Through decades of observation, we have witnessed children exhilarated by the excitement that comes from solving mathematical and logic puzzles, many of them imbued with the spirit of the aforementioned fantastical contexts. We clarify that the idea is not to "make math fun," but simply to reveal what mathematics is: an intellectually stimulating and gratifying adventure.

Stanford University's Ravi Vakil argues that problem solving is "the most stimulating activity" for the mathematically talented — not the standard textbook problems, but rather colorful, even whimsical problems that attract and hold students' attention (Vakil, 1996). During his experience in regular classes in the United States and Latin America, one of the authors has witnessed that many, perhaps most, students also enjoy these challenging intellectual experiences, regardless of the intellectual labels educational systems may put on them.

11.2 Math Circles and Puzzles

Generations of mathematicians and scientists thank Martin Gardner and his ingenious puzzles (Gardner, 1994, 2005) for having aroused their interest in the mathematical sciences. Many also thank Raymond Smullyan (see Smullyan in references at the end of this chapter for a list of his publications), for his extraordinary puzzle-making ability, which renders profound results in mathematical logic intelligible to an educated general audience. Moreover, Russian teachers have used Kordemsky's puzzles (Kordemsky, 1992) to breed generations of young students in their math circles.

The National Association of Math Circles (NAMC), of which one author is an advisory board member, defines math circles as follows:

> Mathematical circles are a form of education enrichment and outreach that bring mathematicians and mathematical scientists into direct contact with pre-college students. These students, and sometimes their teachers, meet with mathematical professionals in an informal setting, after school or on weekends, to work on interesting problems or topics in mathematics. The goal is to get the students excited about mathematics by providing a setting that encourages them to become passionate about mathematics. [...] Athletes have sports teams that deepen their involvement with sports; math circles can play a similar role for kids who like to think about math. One thing all math circles have in common is that the students enjoy learning mathematics, and the circle gives them a social context in which to do so.[1]

Circles bring intriguing and valuable mathematics to students in ways that traditional classroom practices usually do not, partly because circle leaders are not particularly interested in covering a certain amount of material. Rather, they focus on depth of understanding. However, the way many curricula are designed do not allow for the exploration needed to reach significant levels of depth in understanding; curricula tend to focus on quantity of topics taught.

---

[1] http://www.mathcircles.org/content/introduction-to-math-circles.

Certain pedagogical techniques used in circles that are conducive to learning, like creating a relaxed learning atmosphere that is free of anxiety and high-stakes testing, are elements that can be incorporated into the general classroom. More importantly, the social context alluded to above gives circles the sense of community that is key to the success of this approach.

Mathematical and logic puzzles — typical math circle problems — are ideal ways to stimulate creativity in students, as well as effective means to foster their deductive reasoning abilities. Yet, many classrooms around the world lack this vital element. Moreover, the excitement and motivation these problems create can serve as catalysts for developing sound reasoning skills and understanding the validity of an argument — the latter is, in essence, the concept of proof. After all, conjectures are mathematical puzzles that, when solved, become theorems.

Students attending math circles benefit from these educational experiences, but many students are unaware that such excitement even exits in mathematics. Given a chance, they might change their perspective. Hence, it is important that teachers learn to incorporate puzzles into their lessons.

11.3 Gifted Education: A Theory for the General Classroom

A math circle is an ideal setting to fulfill Renzulli's (2012) three objectives for gifted education, as delineated in his Schoolwide Enrichment Model, "a general theory for the development of human potential [...] for creative productivity":

(1) To provide young people with maximum opportunities for self-fulfillment through the development and expression of one or a combination of performance areas where superior potential may be present.

(2) To increase society's reservoir of persons who will help solve the problems of contemporary civilization by becoming producers of knowledge and art rather than mere consumers of existing information.

(3) To provide an organizational and pedagogical framework for strength-based assessment and personalized learning that focus on how students access and make use of information rather than merely on how they accumulate, store, and retrieve it.

Although designed for gifted programs, most people would agree that these goals should become the goals of every classroom.

In terms of strength-based assessment and personalized learning, it is the teacher's responsibility to provide opportunities for students to reveal their talents by eliciting ingenious solutions to challenging problems. Circle leaders are usually "encouraging, even about wrong answers." Yet, leaders are not expected to be "satisfied until there is a rigorous solution" (McCullough and Davis, 2011).

We have also seen that the skills students develop in a circle help them find self-fulfillment through problem solving *and* creation; the latter because they learn to make their own conjectures — future problems to be solved through acts of creativity. Besides, this approach will also subvert the exclusiveness of gifted programs, giving more students access to a high-quality mathematics education.

11.4 Informal Arguments and Puzzles

Informal arguments are what Polya referred to as "plausible reasoning [...] This is the kind of reasoning on which [...] creative work will depend" (Polya, 1954). Fostering this type of reasoning and stimulating students' creative talents are facilitated by identifying problems that can be approached by concrete reenactment. However, "If we wish to talk about mathematics in a way that includes acts of creativity and understanding, then we must be prepared to adopt a different point of view from the one in most books about mathematics and science" (Byers, 2007). This approach emphasizes the dynamic nature of mathematics, which is shown even more dramatically if the students are allowed to discuss their ideas among themselves as they try to solve the problems.

A key element of math circle problems is that they are usually stated in simple and direct language — often deceptively simple — that allows for an easy entry for everyone. Think of Goldbach's conjecture: every even number greater than two is the sum of two prime numbers. Of course, we also want students to be able to *solve* the problems, so we must carefully select or adapt problems suitable for their ability. The idea is that students will be attracted by the problem's apparent simplicity and charm, so that by the time they realize the problem is not as easy as it seemed upon first glance, they are already engaged in trying to solve it and will persist. In order to keep frustration at bay, teachers need to know when to give hints, without giving the problem away and thus depriving students of experiencing a sense of discovery.

Teachers may wonder what happens when students realize that most puzzles posed are "deceptively simple." Will they give up before even trying, knowing that the problem might be difficult? Interestingly, we have noticed that once students learn about this trap, they continue to fall for it — albeit consciously — perhaps because they enjoy the excitement of the possibility of succeeding at a difficult task. In fact, we have witnessed how, even when presented with a problem whose solution has been shown to be impossible — like filling a 4 × 5 checkered board with the five tetrominoes — students continue to play with the problem until they are convinced the task is impossible.

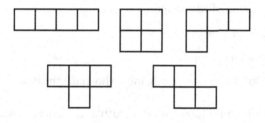

Tetrominoes

Tetrominoes

11.5 Sample Problems

Problems might include some sensory input that can be approached by concrete reenactment, like the following one.

11.5.1 *Baba-Yaga*

"You're not ready to see this," said Baba-Yaga to her 33 students. "Close your eyes," she commanded. All the boys and one third of the girls closed their right eyes; all the girls and one third of the boys closed their left [eyes]. How many students were able to get a peek, despite Baba-Yaga's demand? (Yashchenko, 2013)

This problem appeared in a 2011 Russian competition for sixth and seventh graders (11–13 years). In 2013, we used this problem with a mixed group of second to fifth graders (7–11 years) and with another group, of seventh and eighth graders (12–14 years). Most students really enjoyed this problem, in part because they could reenact it. They began to role-play, telling their peers to close their eyes this way or that way, and so on. To our surprise, one second grader, who had just learned multiplication and had not been taught division, solved the problem after we gave him a very succinct explanation of what "one-third" was.

11.5.2 *Dragon-fighting knights*[2]

In a newly unearthed story of the Knights of the Round Table, three knights come to King Arthur to inform him about the slaying of a dragon.

Gawain: Lancelot killed a dragon.

Lancelot: Tristan killed a dragon.

Tristan: I killed a dragon.

If only one knight told the truth, then who killed the dragon?[3]

Only Gawain could have told the truth and, hence, Lancelot must have killed the dragon. (Why a knight would lie is an interesting moral question!)

Problems such as this are an excellent introduction to logical reasoning, and young students seem to revel in them. Experience informs us that even those who might come into our classroom with resistance to mathematical learning, feel safe attacking these problems, perhaps because they do not see a connection to mathematical thinking. We take advantage of this situation and later guide them into exploring the connections between logical and mathematical thinking with questions like: *Is this mathematics? If it is not, what is it? If it is, how is it so?* You may need to delve more deeply into the subject and ask: *What is mathematics? Is mathematics about numbers? Is mathematics about patterns?* These are open questions worth pondering from an early age.

---

[2] This problem is a modification of one found in Rozhkovskaya (2014).

[3] These truth-tellers and liars puzzles were made famous by Raymond Smullyan.

11.5.3 *Cutting a square into squares*

One of our favorite problems is this: How many squares can you cut a square into? We like this problem because it allows for multiple levels of difficulty. We can ask students whether they can cut a square into 6, 7, or 11 squares, or why we cannot cut it into 2 or 3. At an advanced stage, we may pose the following question: "For what whole numbers can we cut a square into that same number of squares?" (The answer is for any positive integer other than 2, 3, or 5. There is a very elegant and easy way to show how to do this for members of the solution set. Find it!)

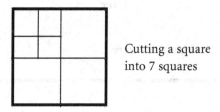

Cutting a square
into 7 squares

We have tried this problem with primary and secondary students, as well as with college students — who, incidentally, were the least receptive and most hostile to the presentation. We let students encounter and artic-ulate difficulty. "What is a square?" we would first ask them, if they had not already inquired. "How do you know the shapes you're cutting are actually squares instead of just rectangles or something else?"

11.5.4 *Cow and farmer*

Here is a problem worth exploring with students of different ages:

> A farmer and a cow are on the same side of a straight-line river. The farmer has to walk to the river, get water in a bucket, and take it to the cow. What's his shortest path? (McCullough and Davis, 2011, p. 3)

The beauty of the solution is that it can be constructed by imagining the reflection of the cow on the river and realizing that it takes the same amount of time to reach the cow as it takes to reach its reflection (provided we could walk on water or swim at the same speed!). Since the way to the reflected cow is a straight line, the farmer should take this path to reach the point at the edge of the river where he will get water, and then head straight to the cow. This is so because the path to the reflected cow and the path to the actual cow from the point the farmer collected water make equal angles of incidence and reflection off the edge of the river, a concept that can be intuitively understood by elementary school students.

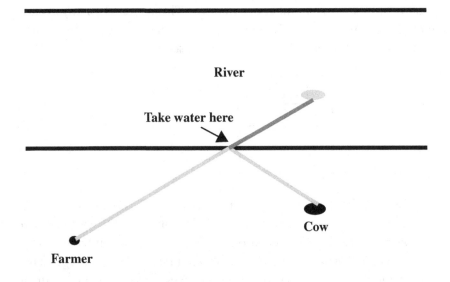

11.6 Puzzles and Proofs

Formal proofs in mathematics are similar to solutions of mathematical puzzles. There are important differences, like the complexity of the technical and historical context of mathematics. Yet, if students learn to provide articulate and detailed responses to carefully selected puzzles, they will be better equipped to master the intricacies of mathematical reasoning. After all, although remembering axioms, definitions, and previous results is necessary to prove theorems, the most important component in internalizing the concept of proof lies in understanding what deductive reasoning is. It is the responsibility of teachers to help students recognize the connection between solving a problem and constructing a proof by demanding detailed explanations of their reasoning, instead of encouraging complacency after finding an answer. As a bonus, the informal setting of puzzles can help teachers identify mathematical talent that may otherwise go unnoticed.

11.7 Incorporating Puzzles into the Curriculum

It is incumbent upon mathematics teachers to provide opportunities for students to reveal their talents by eliciting creative solutions to challenging problems. As a general principle, teachers should strive to present mathematics as an interesting and accessible intellectual game, rather than as a set of rules and procedures of practical necessity. In order to present mathematics as a living field, teachers must not be afraid to digress to explore an idea or comment offered by students or to pose questions for which answers may not be readily accessible.

Puzzles come in all sorts of forms and contexts, and can fit the curriculum of any classroom at any level. Different solutions to the same problem must be welcomed and encouraged. Moreover, teachers should select problems where students can apply similar techniques to those they have used to solve another problem.

11.8 Egalitarian Focus

Bringing the culture of math circles to the math classroom might be a key element for creating excitement in learning mathematics. Again, it would also subvert the exclusiveness of gifted programs, giving more students access to a high-quality mathematics education. Moreover, mathematical and logic puzzles foster an intellectually stimulating environment that will not only help to create a better foundation on which to build the structure of mathematical knowledge, but advance a culture that values mathematical ideas. This should be the goal of any mathematics education agenda.

References

Byers, W. 2007. How Mathematicians Think: Using Ambiguity, Contradiction and Paradox to Create Mathematics. Princeton, N.J.: Princeton University Press.

Gardner, M. 1994. My Best Mathematical and Logic Puzzles. New York: Dover Publications.

Gardner, M. 2005. The Colossal Book of Short Puzzles and Problems. New York, N.Y.: W.W. Norton and Company, Inc.

Kordemsky, B.A. 1992. The Moscow Puzzles: 359 Mathematical Recreations. New York: Dover Publications.

McCullough, E. and Davis, T. 2011. So You're Going to Lead a Math Circle. Retrieved July 1, 2013, from http://geometer.org/mathcircles.

Polya, G. 1954. Mathematics and Plausible Reasoning: Volume 1. Princeton, N.J.: Princeton University Press.

Renzulli, J.S. 2012. "Reexamining the role of gifted education and talent development for the 21st century: A four-part theoretical approach", Gifted Child Quarterly, 56(3), 150–159.

Rozhkovskaya, N. 2014. Math Circles for Elementary School Students. MSRI Mathematical Circles Library. American Mathematical Society, from http://www.ams.org/bookstore-getitem/item=mcl-13.

Smullyan, R.: Works include:

- (1978) What Is the Name of This Book? The Riddle of Dracula and Other Logical Puzzles ISBN 0139550623
- (1979) The Chess Mysteries of Sherlock Holmes ISBN 0394737571
- (1982) The Lady or the Tiger? ISBN 0812921178
- (1982) Alice in Puzzle-Land ISBN 0688007481
- (1985) To Mock a Mockingbird ISBN 0192801422
- (1987) Forever Undecided ISBN 0192801414
- (1992) Satan, Cantor and Infinity, Alfred A. Knopf: New York.
- (2009) Logical Labyrinths ISBN 9781568814438, A K Peters
- (2013) The Godelian Puzzle Book: Puzzles, Paradoxes and Proofs ISBN 0486497054

Vakil, R. 1996. A Mathematical Mosaic: Patterns and Problem Solving. Burlington, Ontario: Brendan Kelly Publishing Inc.

Yashchenko, I. 2013. Invitation to a Mathematics Festival. MSRI Mathematical Circles Library. American Mathematical Society, from http://www.ams.org/bookstore-getitem/item=mcl-12.

CHAPTER 12

TECHNOLOGY: A TOOL FOR ANALYSIS AND INTERPRETATION

12.1 Introduction

Technology — in the form of calculators and computers — is a tool for learning and teaching mathematics (National Council of Teachers of Mathematics (NCTM), 2000; Schwartz, 1999). It is useful in the development of representations of mathematical concepts and facilitates organization and data analysis. It also allows us to shift our focus to interpretation, comprehension, and evaluation by liberating students from having to do tedious or time-consuming computations. Moreover, it helps with diversifying teaching.

We are not suggesting that technology replace intuition and the understanding of numerical computation. What we are suggesting is that technology be given recognition as a useful pedagogical tool. Certainly students must understand how calculations are done, but once that has been learned, technology can free students from having to do manual calculations so they can focus on interpretation, comprehension, and evaluation.

Integrating technology into teaching requires that we revise not only the courses' content, but the strategies we use to teach them. It is imperative to research effective ways of integrating technology in order to find a balance between understanding the operations behind a calculation and using technology to do the calculations (Moore, 1997; Ware and Chastain, 1989).

12.2 Technology Can Support Learning

Consider the following ways in which technology can support learning (Groves, 1994; NCTM, 2000; Rojano, 1996; Schwartz, 1999; Sheets, 1993).

(1) **Development of representations of mathematical concepts**

As we have stated throughout the book, one of the main problems in mathematics teaching is learning without understanding. A way to teach with understanding is to offer representations that allow students to give meaning to concepts. The graphic power of technological tools gives students access to visual models that reinforce their representations. Graphics and simulations can offer intermediate models enabling students to transition from reality to the abstract world of mathematics. In simulations, we can have certain traits of the real situation, but we can vary different factors and see the effects they generate. For instance, we can see the role that different variables play in a real situation and begin to understand the relationship between them. In this way, mathematical formulas take on meaning and situations begin to be understood from mathematical models.

(2) **Interpretation of situations with mathematical models**

Technology broadens the reach of analyzing a situation by mathematical means because it offers a multi-perspective glance. For instance, with dynamic geometry programs students can change, in various ways, geometric shapes and observe their invariant properties; this can foster the understanding of theorems and geometrical properties (for further information see http://www.math.bas.bg/omi/DidMod/Articles/BB-dgs.pdf). Likewise, spreadsheets can give a quick visual overview of what happens as variables change, which is important for understanding what a mathematical model tries to capture.

(3) **Facilitation of date organization and analysis**

Technology allows us to perform arithmetic, algebraic, and statistical calculations with great ease. This frees students from focusing on this mechanical aspect of analysis so they can focus their attention on what is essential for problem solving: interpretation, comprehension, and analysis.

In Chapter 4, while discussing addition and subtraction, we saw how calculators could help with the teaching of those operations. We reiterate, though, that calculators should not replace the learning of algorithms and basic combinations that allow students to understand the structure of our number system and, thus, perform mental arithmetic and estimations. Once they learn the basic combinations and understand the algorithms, calculators can be used to simplify operations with large numbers or many numbers.

This should be the norm in using technology as a teaching tool. First, students must learn the basic concepts without technology — like learning what an average, an integral, or a matrix is. They should also know how to perform basic operations with them. Once they are able to do this, technology can help in situations requiring time-consuming or tedious calculations, some of which might be impossible to do by hand.

(4) **Diversifying teaching**
We must also recognize that learning and the construction of knowledge are not equal for everyone (Gardner, 1999). For several decades, it has been stated that students are different and follow diverse paths in their development. However, this widely accepted principle is often disregarded in curriculum design, with the resulting curricula being homogeneous in form as well as in pedagogical approach. Technology offers an opportunity to present a diversity of scenarios addressing different learning styles and learning rates. For instance, with computerized modules, a teacher may have in the same class different levels of engagement — for example, some students working on addition while others explore multiplication.

12.3 Need to Revise Both Content and Teaching Methods

Integrating technology requires the revision of both the curricular content and the methodology of teaching (Moore, 1997). On the one hand, technology makes obsolete much of the time we spend practicing algorithms with ever more complicated numbers, which can instead be used to focus on problem solving and the analysis of situations. Thus, content should emphasize interpretation and analysis, instead of calculations that can be done with technology. In fact, during curriculum design we should ask which aspects of what we are teaching can be done more efficiently by a computer than by humans. In these areas, we should offer some basic principles that will give students an idea of the task at hand and teach them how to use technology to carry out the calculations. We should also ask which aspects of what we teach are better done by humans than by computers, and focus on those.

In general, computers outperform humans in algorithmic tasks (Levy and Murname, 2004), like doing addition, subtraction, multiplication, and division problems; solving a system of linear equations; or finding an average. Humans, however, outperform computers in tasks like finding patterns, interpreting situations, and problem solving. Teaching should, then, focus on these areas.

Even in the areas where humans outperform computers, computers can be of help. After all, computers were created by humans. Moreover, technology allows us to include interesting problems, like problems in mathematical modeling, which we were not able to include before owing to the complexity of manual calculations. Teaching strategies are also affected by technology. For instance, computers facilitate the testing of students at different times, according to their learning pace, and repeating testing until certain skills are learned. They also offer opportunities to continuously assess students with relative ease.

12.4 Learning to Code

Learning to code, like teaching a child how to program a calculator to find the arithmetic mean of many numbers or to play Conway's Game of Life, can greatly enhance the learning experience. Not only must students learn how an algorithm works, but they need to learn how to organize the steps of the algorithm in order for the computer to understand their commands. This requires developing good reasoning and problem-solving skills. Coding, in fact, might be the greatest way to integrate technology into the curriculum. In the process, students learn valuable skills that will be beneficial in their future.

12.5 Importance of Teachers

Although technology can be very helpful, it is not a panacea. As with every tool, it can be used efficiently or inefficiently; it is the duty of the teacher to make good use of it. In fact, technology is no substitute for a teacher.

Sometimes, when students use technology, it might seem that the teacher is unnecessary. This is incorrect. The teacher is the one deciding the teaching strategy in which technology is to be integrated. In fact, what is essential for teaching is not technology itself, but integrating that technology in the right way. If we choose an active, constructive, or imaginative approach, technology can be of great use; however, if we choose a boring, passive, or repetitive approach, it can be detrimental. In the latter case, technology could at first be good to break the routine, but once it becomes part of the routine, it will be as boring as the class without it. This reminds us of a question posed to a family member after her computer lab class. "What did you do today?", we asked. "The same as yesterday," she said, unenthusiastically.

As we will discuss in the next chapter, observing how students work is essential to improve our teaching strategies. In the case of technology, for instance, while students are using technology, be it calculators or computers, teachers have an opportunity to observe their reasoning. These observations, in turn, can enrich teaching by helping us to find alternative ways to incentivize learning.

References

Gardner, H. 1999. Intelligence Reframed. New York: Basic Books.

Groves, S. 1994. "Calculators: A learning environment to promote number sense". Paper presented at the Annual Convention of the American Educational Research Association, New Orleans.

Levy, F. and Murname, R.J. 2004. The New Division of Labor: How Computers are Creating the Next Job Market. Princeton, N.J.: Princeton University Press.

Moore, D.S. 1997. "New pedagogy and new content: The case of statistics", International Statistical Review, 65(2), 123–165.

National Council of Teachers of Mathematics. 2000. Principles and Standards for School Mathematics. Reston, Va.: National Council of Teachers of Mathematics.

Rojano, T. 1996. "Developing algebraic aspects of problem solving within a spreadsheet environment", in Bednarz, N., Kieran, C. and Lee, L. (eds.), Approaches to Algebra: Perspectives for Research and Teaching. Dordrecht: Kluwer Academic Publishers.

Schwartz, J.L. 1999. "Can technology help us make the mathematics curriculum intellectually stimulating and socially responsible?", International Journal of Computers for Mathematical Learning, 4, 99–119.

Sheets, C. 1993. "Effects of computer learning and problem-solving tools on the development of secondary school students' understanding of mathematical functions", Ph.D. dissertation, University of Maryland College Park.

Ware, M.E. and Chastain, J.D. 1989. "Computer assisted statistical analysis: A teaching innovation?" Teaching of Psychology, 16, 222–227.

CHAPTER 13

ASSESSMENT

13.1 Introduction

The learning vision we have presented in this book requires new ways of assessment (van den Heuvel-Panhuizen, 1996). For instance, if we understand that students construct their knowledge from what they already know, it is essential to constantly assess what they know and create experiences that will foster their intellectual growth. That is why assessment and teaching must go hand in hand. Assessment is an integral part of the teaching–learning process, from planning to an analysis of its effectiveness. In fact, the main objective of assessment should be to improve teaching (Kulm, 1994).

13.2 First Principle: The Main Objective of Assessment Should Be to Improve Teaching

There are various ways to assess teaching, as well as what students know and what they can learn (Clarke, 1998; National Council of Teachers of Mathematics, 1995; Romberg, 1994). Yet, in all of them it is essential to observe what students do and listen to what they say. Moreover, we should try to understand the logic of their reasoning and explanations. In fact, one of the challenges of assessment is being able to understand what students are thinking. This is not an easy task since often the students' logic is different from that of teachers. It is very easy to listen to and interpret from our own perspective, and from there jump to correct what we perceive as an error, when in reality, in the students' logic, the answer is not incorrect.

Knowing the trajectory of the learning of concepts gives us information about the stages through which students go, and that, in turn, helps us to interpret their reasoning. In this sense, something we may consider a mistake — at whatever stage students are — could reflect sound reasoning. For instance, we once found the following two answers to this problem:

Find the area of this shape.

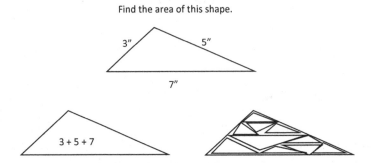

One child added all the sides, and the other made small triangles and added them. Although neither obtained the expected answer, the first one made an error by confusing area with perimeter, whereas the second one has an idea of what area is and how to measure it. In the first case, it is necessary to point out the confusion; in the second case, we use what the student knows to show the need to standardize units of measurement.

This example illustrates the importance of analyzing students' responses and thinking about their reasoning. To advance teaching, we can learn from answers showing a level of understanding as well as from students' mistakes (Lamon and Lesh, 1992).

13.2.1 *Importance of analyzing errors*

Errors are a natural part of learning; it is impossible to learn without making mistakes. Moreover, mistakes are a source of learning for both students and teachers (Clements, 1980). That is why it is imperative to adjust our attitude toward them. To start, we can distinguish between three types of errors: communication errors, interpretation errors, and real errors.

Communication errors

Communication errors are those in which students' thinking and analysis are correct, but there is a problem with language or communication. For instance, consider the following exchange after seeing the picture shown below.

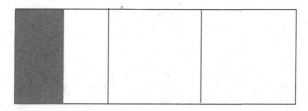

Teacher: What is the shaded region?

Student: Half.

Teacher: Half? Don't you see that it can't be half of the shape?

Student: But it's one-half of one-third.

Oftentimes we have this type of situation, in which there is a communication issue. That is why it is important to ask students to explain and justify their answers. Sometimes, responses that are apparently incorrect are correct, whereas some that seem correct might be the result of flawed reasoning. We can assess the situation by asking students to explain their answers.

Interpretation errors

Interpretation errors are due to students performing the wrong task. That is, the teacher gives some instructions, but students interpret them incorrectly. Under this wrong interpretation the students have worked well, but from the point of view of the teacher they have not.

Real errors

Errors are a natural part of learning. In fact, errors are a useful tool to diagnose how students understand certain concepts and tasks. Consider the following addition and subtraction mistakes we have found. Try to identify the nature of the mistake.

(1) **Addition**

Error 1

27	58	76
+12	+24	+34
39	712	1010

Error 2

38	56	76
+29	+34	+35
121	81	111

Error 3

34	85	78
+72	+96	+94
07	712	613

In the first case, the student added each column without regrouping; in the second one, the student writes the tens and carries the ones; in the third case, the student adds from left to right.

(2) **Subtraction**

Error 1

38	50	40
− 12	− 12	− 17
26	42	37

Error 2

54	47	85
− 27	− 34	− 49
33	13	44

In the first case, when encountering a 0, the student selectively extends commutativity to subtraction:

$$2 - 0 = 0 - 2 = 2.$$

In the second case, the student always subtracts the smaller from the larger number within a column.

Notice that these errors are due to a misconception or an incorrect generalization of a concept. Nevertheless, we realize that there are several patterns. Research on students' errors shows that there are a small number of misconceptions revolving around an even smaller number of mistakes. Hence, all students — especially the struggling, timid type — can benefit from a discussion of the errors found in class.

From the point of view of cognitive development, any lesson, regardless of how good it is, will be incomplete; it is impossible to encompass every possible case of a rule or principle. In fact, teaching assumes that students will infer new knowledge from what they have learned. In this process, errors are bound to arise. As we saw in the second subtraction mistake above — when the student generalized the law of commutativity and extended it to subtraction — many errors make much sense. In fact, some mistake show a deeper level of reasoning and understanding than some correct answers; therefore, we reiterate the need to engage students in discussing their responses, whether they are right or wrong. Everybody, including the teacher, potentially benefits from such exchanges.

13.2.2 *Importance of questions*

We should pay as much attention to students' questions as we do to their answers. We should carefully listen to their queries, and sometimes ask them to elaborate, especially if we think the question could lead to a better understanding of the subject at hand. Even when this does not happen, their questions might still shed light on their reasoning, which is always valuable for a teacher.

13.2.3 *Observation: Source of reflection for teachers*

When observing students we ought to be aware that observation and reflection are two-way processes. This means that we observe students working on activities we create, but also observe our actions in order to adjust and improve the educational activities we develop. Thus, observation is an integral part of teaching to assess if, what, and how students learn. This implies a holistic approach to observation, requiring that we be mindful not only of the concepts we are teaching, but of students' attitudes, their problem-solving strategies, their approaches to problems, and their ability for collaboration. For this to happen, we must give students opportunities to ask questions and support their answers, ideas, or strategies.

The following is a partial list of ways in which observation might take place:

- Passively observe what students do.
- Introspection: Ask students to share their thoughts out loud.
- Retrospection: Ask students to describe how they solved a problem or performed a task.
- Continuous inquiry: Repeat a question in a different way or ask follow-up questions to better understand what students are doing. **When asking a question, make sure to give students time to answer.** A good strategy is to count to ten before asking another question or rewording the same question.
- Reflection upon someone else's work: Ask students to analyze and explain the reasoning of another student or how their strategy differs from others.
- Problem variations: Adjust problems in terms of context and difficulty and observe how student involvement changes.
- Helping students: As students work on a problem, we can provide hints that allow them to discover new paths and strategies. Vygotsky used to say that what students create is as important as what they learn (Vygotsky, 1987).

13.3 Second Principle: Assessment Should Allow Students to Show What They are Capable of Doing, Instead of Emphasizing What They Do Not Know

Assessment should "look back and forward" (van den Heuvel-Panhuizen, 1996). While looking back, assessment tells us what students have learned; this gives us criteria by which to decide what we should teach. Furthermore, assessment should also look forward and see what students are capable of doing; tests and exercises should foster this approach. This emphasis is in line with our thesis, which posits that students construct new knowledge from what they already know. This allows us to set high and realistic goals. It also helps us to adjust the learning process according to group and individual needs. In fact, learning sequences that follow the logic of learning offer a wide range of possibilities for individualized learning, since each student can progress differently.

Tests normally used do not provide sufficient information about the students' thinking processes, especially about what they are capable of doing. It is necessary to broaden the problems we introduce to let them inform us about how students are thinking; the difficulties they are facing; and the knowledge, strategies, and tools available to them. Let us analyze some misconceptions about tests that limit their potentiality to foster learning.

13.3.1 *Misconceptions about tests*

(1) **There is only one way to solve a math problem**
Most math problems can be solved in several ways. The ways in which students solve them offer valuable information about their knowledge and competence. A way to see which strategies they have used is to assign space for computation. For example:

ADD	Show your work in the space below.
13	
25	
27	
25	

While doing this exercise with a group of students, we found the following strategies in the space provided for computation:

Child 1	Child 2	Child 3		Child 4	
63	13	27	25	10	3
+27	25	+13	+25	20	5
90	27	40	50	20	7
	+25			+20	+5
	90	40 + 50 = 90		70	20

This work lets us see diverse strategies. Sometimes, this diversity also shows a gradation of understanding. This allows the teacher to serve each child according to the individual needs of that particular student. It also offers information about what the student is capable of doing. Discussing students' strategies can help them learn more efficient strategies from one another.

(2) **A math problem has a unique solution**

Open problems, or problems with more than one solution, allow us to better understand student strategies. For instance:

The Rivera family went to a music concert. The family consists of Dad, Mom, ten-year-old Luis, and eight-year-old Mariana. If they paid $50 in tickets for the entire family, how much did each ticket cost?

Show your work here.

What possible solutions does this problem have? There are different criteria we can use to decide the price of the tickets. For example, we can say that each adult paid $15 and each child $10. We can also divide the cost evenly. This implies that we must understand decimals, since the answer is $12.50 per ticket. In fact, there are infinitely many solutions to this problem. Observing the strategies students choose allows us to see their different aptitude levels.

In this problem there might be students who do not know what to do. While observing students during problem-solving activities, we should not limit ourselves to what they can answer on their own. We should pose questions that will help them to understand and learn (Brown and Walter, 1983). Knowing what students know is as important as assessing what they can learn.

More open problems: There are problems in which we do not have all the information needed to solve them. Part of the task is to search for that information and make some decisions about it. For example:

There is a 500-pound bear at the zoo. How many children from your class will it take to equal the bear's weight?

500-lb bear

Answer:

This problem has several solutions. To start, the answers will vary from grade to grade as, on average, older children are heavier than younger ones. Also, the number of students will depend on how they are chosen. For example, if we choose the heaviest ones, we will need fewer students than if we choose the lightest ones. If we choose them at random, the number would be closer to the average for the entire class. Hence, a problem like this can elicit interesting responses and conversations, some of which might lead us to the concept of the arithmetic mean.

(3) **Mathematical knowledge that has not been taught cannot be assessed**
Another misconception that ought to change is that we cannot assess concepts that have not been taught. In fact, assessing such concepts might give us ideas about the informal knowledge that students have, and from there, allow us to construct a new concept.

Example: Prior to introducing division problems, we may use the following problem.

Number of peaches	Number of boxes	Peaches per box	Number of peaches	Number of boxes	Peaches per box
36	6		54	9	
48	6		56	8	
18	6		63	7	
54	6		49	7	
24	6		45	5	

The different strategies that students use to solve this problem might give us an idea about where we should start when teaching division. Consider the following strategies used by students when faced with this problem. Let us focus on the case with 24 peaches in 6 boxes.

Student 1	Student 2	Student 3
Draws 6 boxes and starts to distribute the 24 peaches.	Starts adding different numbers until noticing that 4 + 4 + 4 + 4 + 4 + 4 = 24.	6 x 4 = 24. Four peaches per box.

13.4 Third Principle: Every Objective Should Be Assessed

If we cover different objectives in class — like problem solving, mathematical reasoning, communication, making connections — but we do not assess them in tests, assignments, or otherwise, then students may get the wrong idea that what we assess is what is important while the rest is optional. We should, then, assess the different course objectives. Consider the following example.

(1) A rock band is planning its next concert. They want a rectangular stage covering 1000 square feet and surrounded by security rope. Draw possible rectangles representing an area of 1000 square feet. How many feet of rope would you need to cover each one? Which design would you choose? Why?

In this problem, students must **solve a problem**, and they can do this with different strategies. They must **reason** to decide which alternative is the best. Finally, they must **communicate** their answers.

(2) Identify all similarities and differences you can find in these two shapes. Then, draw a shape that looks more like A than B. Explain your choice.

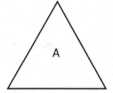

In this problem, students must **reason** to identify similarities and differences between both shapes. Then, they must **make connections** between properties to choose a shape that resembles A more than B. Finally, they have to explain, i.e. **communicate**, their responses.

13.5 Fourth Principle: When Using Assessment as a Tool to Grade Students, the Score per Problem Should Not be "All or Nothing"

While grading exam problems, it is pertinent to analyze all possible levels for an answer and, accordingly, assign a score to each level. For instance, in the concert stage problem above, students offered the following answers.

Answer 1

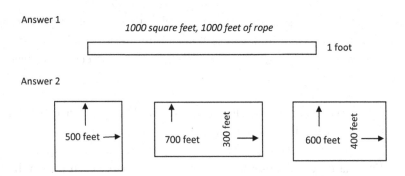

Answer 2

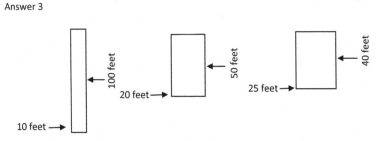

*I need 2,000 feet of rope in all of them. I'd choose 700 feet
long and 300 feet wide because there's more space.*

Answer 3

*In the 10 x 100 I need 220 feet of rope, in the 20 x 50 I need 140 feet of rope and in the 25 x
40 I need 130 feet of rope. I'd choose 25 x 40 because it requires less rope to round the stage.*

— The first student has almost no notion of area or perimeter. He drew
   only one rectangle and wrote that he had 1,000 square feet and that he
   needed 1000 feet of rope.

— The second student does not understand the concept of area very well,
   and apparently adds width and length instead of multiplying. Yet, she
   understands and knows how to find the perimeter. It would be good
   to ask her why the area of those rectangles is 1,000 square feet.

— The third student seems understand both concepts well and also
   shows sound reasoning.

These three answers are quite different. Since there are distinct levels
of understanding, we should take this into consideration while grading
each answer. To organize how to score them, we develop a rubric.

13.5.1 *Rubrics*

A rubric is a strategy to organize the scoring of an assignment, which depends on the type of task. We show an example of a rubric for the stage problem above. This problem requires three steps: understanding the problem, solving it, and explaining the situation. In fact, in many mathematical problems these three elements play a role.

(1) **Understanding the problem**
   - 4 points — Thorough understanding.
   - 3 points — Some difficulty understanding the problem.
   - 2 points — Difficulty in understanding the problem.
   - 1 point — Poor understanding.

(2) **Problem-solving process**
   - 4 points — Correct.
   - 3 points — Almost correct.
   - 2 points — Attempt in the right direction.
   - 1 point — Trying.

(3) **Explanation**
   - 4 points — Complete.
   - 3 points — Almost correct.
   - 2 points — Incomplete.
   - 1 point — Poor.

As an example of how to use the rubric, we score the three answers given above.

Answer 1:
   - Understanding the problem: 1 point.
   - Problem-solving process: 1 point.
   - Explanation: 0 points.
   - Total: 2 points.

Answer 2:
   - Understanding the problem: 3 points.
   - Problem-solving process: 2 points.

- Explanation: 2 points.
- Total: 7 points.

Notice these answers are both incorrect, but show different degrees of understanding; hence, the scores are different.

Answer 3:

- Understanding the problem: 4 point.
- Problem-solving process: 4 point.
- Explanation: 4 points.
- Total: 12 points.

13.6 Conclusion

We have discussed several assessment principles separately. In the teaching process, however, it is often imperative to integrate several or all of them. For instance, while grading, we can realize whether the teaching of a particular subject needs further elaboration. Allowing the assessment process to inform us about what students are capable of can generate ideas about how to introduce a certain concept. Thus, we should keep all of these principles in mind during assessment.

References

Brown, S.I. and Walter, M. 1983. The Art of Problem Posing. Hillsdale, N.J.: Lawrence Erlbaum Associates.

Clarke, D. 1988. Assessment Alternatives in Mathematics. Canberra: Curriculum Development Centre.

Clements, M.A. 1980. "Analyzing children's errors in written mathematical tasks", Educational Studies in Mathematics, 11, 1–21.

Kulm, G. 1994. Mathematics Assessment. San Francisco: Jossey-Bass Publishers.

Lamon, S.J. and Lesh, R. 1992. "Interpreting responses to problems with several levels and types of correct answers", in Lesh, R. and Lamon, S. J. Assessment of Authentic Performance in School Mathematics. Washington, D.C.: AAAS Press, pp. 319–342.

National Council of Teachers of Mathematics. 1995. Assessment Standards for School Mathematics. Reston, Va.: National Council of Teachers of Mathematics.

Romberg, T.A. 1994. Reform in School Mathematics and Authentic Assessment. New York: SUNY Press.

Van den Heuvel-Panhuizen, M. 1996. Assessment and Realistic Mathematics Education. Utrecht, the Netherlands: Freudenthal Institute.

Vygotsky, L. 1978. Mind and Society. Cambridge, M.A.: Harvard University Press.

CHAPTER 14

CONCLUDING REMARKS

14.1 Introduction

In this book we have analyzed the way in which mathematics is learned. We have emphasized the cognitive development of the mathematical concepts that are learned in elementary school. By understanding the way in which students learn, the teacher can organize educational activities in ways that support learning. The discussion about cognitive development should also promote reflection on the curriculum. A curriculum is more than a collection of activities; it must be coherent, focused on important mathematical concepts, and well sequenced among grades so that mathematical concepts can be developed according to the logic of learning.

14.2 Need to Rethink the Curriculum

Experience shows that a great number of students do not learn with understanding the mathematical concepts that are taught in elementary school. We think that a great deal of this difficulty is due to the fact that these are not taught for understanding. As we have argued throughout the book, to learn with meaning we need to follow the logic of learning. If we analyze many curricula in place around the hemisphere, we will see that conceptual development does not always follow this logic. For instance, research indicates that learning certain topics, like fractions, takes longer and is more complex than curricular guidelines suggest. In this sense we

265

propose that, along with assessing teaching following the logic of learning, the order of the curricular concepts be evaluated. Moreover, only if it is feasible and advisable do we suggest keeping all topics currently included at the elementary level. If a topic must be removed, it must be done after careful deliberation supported by research.

However, we wonder if it is more reasonable, instead of covering so much material, to concentrate on teaching for understanding the basic concepts. In fact, if we analyze the elementary level content plan from the Netherlands' Freudenthal Institute, we note that they limit the range of topics, focusing on the essential concepts, as shown in the following table.

Key Goals for Dutch Primary School Students in Mathematics.

		By the end of primary school, the students
General abilities	1	Can count forward and backward with changing units
	2	Can do addition tables and multiplication tables up to ten
	3	Can do easy mental arithmetic problems in a quick way, with insight into the operations
	4	Can estimate by determining the answer globally, also with fractions and decimals
	5	Have insight into the structure of whole numbers and the place–value system of decimals
	6	Can use the calculator with insight
	7	Can convert simple problems that are not presented in a mathematical way into mathematical problems
Written algorithms	8	Can apply the standard algorithms, or variations of these, to the basic operations of addition, subtraction, multiplication, and division in simple context situations
Ratio and percentage	9	Can compare ratios and percentages
	10	Can do simple problems on ratio
	11	Have an understanding of the concept of percentage and can carry out practical calculations with percentages presented in simple context situations
	12	Understand the relation between ratios, fractions, and decimals

(Continued)

(Continued)

		By the end of primary school, the students
Fractions	13	Know that fractions and decimals can represent several meanings
	14	Can locate fractions and decimals on a number line and can convert fractions into decimals; also with the help of a calculator
	15	Can compare, add, subtract, divide, and multiply simple fractions in simple context situations by means of models
Measurement	16	Can read the time and calculate time intervals; also with the help of a calendar
	17	Can do calculations with money in daily life context situations
	18	Have an insight into the relation between the most important quantities and corresponding units of measurement
	19	Know the current units of measurement for length, area, volume, time, speed, weight, and temperature, and can apply these in simple context situations
	20	Can read simple tables and diagrams, and produce them based on own investigations of simple context situations
Geometry	21	Have mastered some basic concepts with which they can organize and describe a space in a geometrical way
	22	Can reason geometrically using building blocks, ground plans, maps, pictures, and data about positioning, direction, distance, and scale
	23	Can explain shadow images, can compound shapes, and can devise and identify cut-outs of regular objects

If we compare these goals with the old National Council of Teachers of Mathematics *Standards and Principles*, or even with the shortened Common Core State Standards, we notice that the Dutch focus on fewer topics and basic competencies, but with an emphasis on understanding (National Council of Teachers of Mathematics, 2000).

In fact, there are topics that are not covered — for instance, probability, elements of number theory, and the language and operations of sets. Moreover, the Dutch do not require students to learn the algorithms for basic operations with fractions, but only that they understand their meaning. If we ponder this last point, we realize that all basic operations with fractions can be done with decimals. For those students who will not study or need higher mathematics, being able to do these operations with decimals should suffice; students needing algebra for their future academic plans could learn the operations with fractions later, so that not all students are forced to go through this process. We understand that it is more important to learn basic concepts with understanding than to cover more material without it.

Furthermore, besides revising content, it is imperative to revise the order and continuity of curricular development. Once there is an agreement upon a topic distribution by grade level that fosters the construction of mathematical concepts, teamwork among teachers is essential so that they ensure continuity in students' development throughout grade levels. Each team of teachers ought to make whatever adjustments are needed to suit their circumstances.

Hence, the mathematics curriculum must provide a roadmap that supports teachers in promoting a conceptual development in students that is rich in depth and abstraction. For each concept, teachers must know what students were supposed to learn in the previous grade and what they will cover in the next. For instance, in first grade (six to seven years), students begin their acquaintance with unit fractions up to 1/4. In second grade (seven to eight years), they broaden their knowledge of unity fractions up to 1/10 and begin to analyze, by way of illustrations, fraction equivalencies. In third grade (eight to nine years), they begin working on other fractions by way of unit fractions: 3/4 = 1/4 + 1/4 + 1/4. They also see these fractions in different contexts. Thus, students delve more deeply into several concepts. In Chapter 6 we have addressed and broadened the cognitive development of these concepts.

If we do not achieve this articulation among different grade levels, then we have duplication in teaching and endless remedial sessions. A well-articulated curriculum offers teachers guidelines about fundamental topics and the depth that is expected to be reached at each level.

14.3 Need to Create Learning Environments for Teachers

Regardless of how perfect a curriculum is, it could never encompass all the richness and diversity that are found in a classroom. Hence, along with a coherent curriculum, it is necessary to create a school environment that fosters reflection and the continuous learning of teachers. In fact, it has been said (Shon, 1991) that much of the information required to be productive in teaching emerges from practice. In practice, teachers must merge their content knowledge with learning theories and development, culture, and pedagogical principles. Likewise, they must address student diversity. There are no easy recipes for integrating all these elements. In fact, learning how to coordinate these elements is refined only with practice.

Effective teaching requires that teachers experience a continuous process of improving their teaching as a result of what they learn in practice. To learn from their practice, teachers must observe students and listen carefully to their ideas, questions, and explanations. What teachers learn about how students are learning will serve as a starting point to help students in their construction of mathematical knowledge. In this process, they must provide a variety of opportunities that challenge students of diverse mathematical abilities.

This reflection and analysis process is an individual task for teachers. Nevertheless, it has been shown that this experience is enriched by sharing it with colleagues (Ma, 1999). Ma compared the teaching of elementary mathematics in China and the United States, and found that one of the elements that contribute to more effective teaching in China is the work in discussion groups among teachers. Collaboration among peers who regularly observe, analyze, and discuss each other's classes — along with studying the way students think, as well as sharing their observations and ideas — becomes a powerful professional development tool. Yet, in the United States this is rarely used (Stigler and Hiebert, 1999). It is pertinent to rethink the way of organizing teachers' schedules to allow for these discussions.

14.4 Students can Learn Mathematics with Understanding and Interest

Children have a natural interest in understanding what they learn. From an early age they show interest in mathematical concepts (Gelman and

Gallistel, 1978; Resnick, 1989). Throughout their daily experiences, they acquire a series of informal ideas about mathematics. The type of experience that teachers offer is crucial for these experiences to become mathematical knowledge.

14.5 Final Remarks

In this book we have shown the cognitive development of the concepts taught in elementary school as an aid to educators in their teaching preparation. Yet, the learning of these concepts following their cognitive development must take place in a discourse community (Silver and Smith, 1997) that promotes a genuine interest in problem solving (Schoenfeld, 1985); making conjectures; and in sharing, discussing, and arguing those conjectures with their peers. This environment must be alive and support students' participation in the process of using mathematics with understanding to comprehend situations that are of interest to them. It must also allow for the development of basic skills and understanding of mathematical language, so students can enter that wonderful world of mathematical creation. Thus, it is imperative to seek interesting contexts for the learning of mathematics during the process of revising the curriculum and teaching.

References

Gelman, R. and Gallistel, C.R. 1978. The Child's Understanding of Number. Cambridge, Mass.: Harvard University Press.

Ma, L. 1999. Knowing and Teaching Elementary Mathematics: Teachers' Understanding of Fundamental Mathematics in China and the United States. Hilsdale, N.J.: Lawrence Erlbaum Associates.

National Council of Teachers of Mathematics. 2000. Principles and Standards for School Mathematics. Reston: National Council of Teachers of Mathematics. http://www.corestandards.org/Math/. http://www.uu.nl/en/research/freudenthal-institute.

Resnick, L.B. 1989. "Developing mathematical knowledge", American Psychologist, 44(2), 162–169.

Schoenfeld, A.H. 1985. Mathematical Problem Solving. New York: Academic Press.

Shon, D.A. 1991. Educating the Reflective Practitioner. San Francisco: Jossey-Bass.

Silver, E.A. and Smith, M.S. 1997. "Implementing reform in the mathematics classroom: Creating mathematical discourse communities", in Reform in Math and Science Education: Issues for Teachers. Columbus, Ohio: Eisenhower National Clearinghouse for Mathematics and Science Education. CD-ROM.

Stigler, J.W. and Hiebert, J. 1999. The Teaching Gap: Best Ideas from the World's Teachers for Improving Education in the Classroom. New York: The Free Press.

INDEX